How To Build
BIG INCH
CHEVY
SMALL BLOCKS

SAN DIEGO PUBLIC LIBRARY
LOGAN HEIGHTS

APR 0 6 2001

3 1336 05373 8945

Published by CarTech, Inc.
11605 Kost Dam Road
North Branch, MN 55056 U.S.A.
United States of America
Tel: 800-551-4754, Fax 651-583-2023

Published and Distributed under
license from Petersen Publishing Company,
L.L.C. © 1999 and
® Petersen Publishing Company, L.L.C.
All Rights Reserved

No part of this book
may be reproduced or transmitted
in any form or by any means, electronic
or mechanical including photocopying,
recording or by any information storage
and retrieval system, without written
permission from the Publisher.
All rights reserved

ISBN 1-884089-43-7

Book Trade Distribution by
Voyageur Press, Inc.
123 North Second Street
P.O. Box 338
Stillwater, MN 55082 U.S.A.
Tel: 651-430-2210, Fax 651-430-2211

Distributed in England by
Brooklands Books Ltd
P.O. Box 146, Cobham
Surrey KT11 1LG
England
Tel: 01932 865051, Fax 01932 868803

Distributed in Australia by
Brooklands Books Ltd
1/81 Darley St, P.O. Box 199
Mona Vale, NSW 2103
Australia
Tel: 02 9997 8428, Fax 02 9979 5799

Printed in Hong Kong

CONTENTS

HOW TO BUILD A STROKER SMALL-BLOCK THAT WILL TAKE THE SQUEEZE OF NITROUS 4
THE SMALL-BLOCK CHEVY PUMPS OUT HORSEPOWER AND TORQUE LIKE A BIG-BLOCK 8
INCHING UP ON PERFORMANCE WITH A 400-CID SMALL-BLOCK FOR F-BODIES 12
HOW TO BUILD A STREET-WISE 383 SMALL-BLOCK V8 THAT DELIVERS 450LBS-FT OF TORQUE .. 14
AT LARGE - BUILDING A 454-CUBIC-INCH SMALL-BLOCK 18
THE IMMACULATE DECEPTION - CAMARO GETS A STRONGARM POWERPLANT 22
THE SMALL-BLOCK TAKES ON NEW STATUS BY GROWING AS LARGE AS 482 CUBIC INCHES ... 28
BUILDING AN ALL-ALUMINUM, 520HP SMALL-BLOCK FOR A RADICAL STREET ROD 32
THERE ARE STILL WAYS TO HARNESS A 400 INCHES BUST 36
BUILDING A BIG-CUBIC-INCH BRACKET RACE ENGINE DOESN'T HAVE TO COST BIG BUCKS ... 40
LOOKING INSIDE A 200MPH CHEVY 44
BUILDING THE SMALL-BLOCK CHEVY INTO A FIRE-BREATHING MONSTER OF 440 ,454 OR 482 .. 49
IN A STREET ENGINE, TORQUE IS WHAT YOU WANT AND THIS SMALL-BLOCK CHEVY DELIVERS .. 52
REALITY MOUSE - CAN A SMALL-BLOCK CHEVY MAKE 450 HORSEPOWER ON 92 OCTANE? ... 57
MAXI-MOUSE - STRIP-FLOGGING PAW'S 383 STREET SMALL-BLOCK 62
406-INCH BRACKET SHORT-BLOCK - RELIABILITY AT A REASONABLE PRICE 67
SCRATCH-BUILT 377 - DO YOU KNOW WHAT IT TAKES TO BUILD YOUR FIRST ENGINE - PART 1 .. 70
SCRATCH-BUILT 377 - THE MACHINE WORK ON A HIGH-REVVING SMALL-BLOCK - PART 2 76
SCRATCH-BUILT 377 - "HOT STREET" ENGINE GETS CYLINDER HEADS - PART 3 80
SCRATCH-BUILT 377 - FITS PERFECTLY INTO THE NEW HOT ROD DRAG RACING SERIES - PART 4 .. 84
SCRATCH-BUILT 377 - HOT STREET POWER TEST, WE JUST MISS 600HP - PART 5 88
THE GREAT 408 - BUILD A 600HP RACE SMALL-BLOCK 91
SMALL-BLOCK OUTER LIMITS - MACHINING THE BLOCK - PART 1 94
SMALL-BLOCK OUTER LIMITS - EXPLORING THE ASSEMBLY ON A 427-CUBIC-INCH - PART 2 ... 98
SMALL-BLOCK OUTER LIMITS - THE FINAL ASSEMBLY OF THE MEGA-CUBE - PART 3 102
UNLOCKING THE POWER OF THE 427 - TESTING THE ORIGINAL COMBINATION - ROUND 1 ... 107
UNLOCKING THE POWER OF THE 427 - REACHING THE OUTER LIMITS - ROUND 2 111
SUPER MOUSE 402 CHEVY ABLE TO LEAP TALL BIG-BLOCKS IN A SINGLE BOUND! 120
BRODIX STREET HEADS AND A LUNATI 408 MOUSE MOTOR TEAM UP TO BUILD OVER 475LBS-FT .. 125
JUNKER TO STROKER - CHEVY 383 STREET TORQUER 130

SQUEEZE PLAY

PHOTOS BY ED TAYLOR

How to Build a Stroker Small-Block V-8 That Will Take the Squeeze of Nitrous

By John Kiewicz

Nitrous oxide is a wonderful thing. As one of the best bang-for-the-buck aftermarket power-adding parts available, nitrous oxide injection is also relatively simple to install. Usually within a day's time you can have a new nitrous system installed and ready for use. With the press of a button, your engine's output can easily be boosted by 50-150 hp. However, if even more horsepower is what you crave, then you can opt for more-aggressive-performing nitrous kits that deliver a 200-500hp boost.

Although opting for a big-time 200-500hp nitrous kit seems enticing, you must carefully note that with that much of an added horsepower hit you must have a properly fortified engine to take the added stress. Building such a heavy-duty nitrous engine is the focus of this story.

Steve Porretta was in the midst of building a serious street/strip '64 Nova SS. To make sure that his Nova was a serious player, Porretta wanted to incorporate two major items: big cubes from a stroker small-block V-8, and a serious nitrous hit from a new Nitrous Works port-injected nitrous system.

Building a stroker V-8 isn't too difficult these days, thanks to lots of great aftermarket parts. However, building such a stroker engine with the thought that it will frequently see big doses of nitrous does complicate the matter of making it bulletproof. Here we give some insight on specific parts and assembly techniques that you should consider when building a nitrous-fed engine. Besides showing heavy-duty parts, we also highlight special parts that should be considered, as well as specific items to check when assembling the combo. After Porretta's stroker engine was built, it was dyno-tested to determine its power output with *and* without the nitrous squeeze.

1 To ensure this engine stays together when nitrous oxide is used, a Summit crank kit (PN SES-3-51-55-453) was utilized. The kit includes a special, fully balanced forged-steel crankshaft and I-beam rods that are constructed of 4340 aircraft-quality steel. TRW forged flat-top 10:1 pistons and Speed-Pro rings are included, as well as Clevite 77 H-series main and rod bearings.

2 For this engine buildup, a Summit stroker small-block (PN SES-3-60-04C401) was used. The block has been fully CNC-machined, bored, and honed using a torque plate, and the deck height has been machined exactly parallel to and equidistant from the main bearing bore centerline. Heavy-duty splayed four-bolt main caps have been fitted to the Summit block to ensure that the crank is securely contained, as the nitrous system engages violently.

3 To provide proper clearance for the longer-than-stock-stroke crankshaft, the engine block must have the lower edges of its cylinder bores machined. For this engine buildup, a Summit stroker block that incorporated all of the proper clearancing was used.

4 Even after clearancing the cylinder block, with the crankshaft and rods installed, notice how close the crank/rod combo comes to the edge of the block's oil pan rail. Obviously, proper clearance is mandatory to prevent parts breakage.

5 Although the Summit crank kit is fully machined and balanced, we double-checked all of its specs to ensure that all measurements were correct. For this nitrous 383, the main bearing clearance was 0.0027 inch and the rod bearing clearance was 0.0025 inch.

6 When nitrous oxide is used, in many instances increased crankcase pressure will occur. Added crankcase pressure will often cause oil and/or blow-by to be forced past the rear main seal. To help prevent this, the Fel-Pro rear main seal was slightly cocked, which eliminates the traditional block and seal parting line from being lined up, which can allow blow-by to escape with more ease.

7 To contain the extremely high cylinder pressures encountered during nitrous use, Speed-Pro Plasma-Moly (5/64-inch top ring, 5/64-inch second ring, 3/16-inch oil ring) file-fit rings were used. Due to the added cylinder pressure and heat experienced with nitrous use, special ring endgaps are required. The rings were installed square in the bore, then carefully checked with a feeler gauge.

8 The Speed-Pro file-fit ring endgaps needed to be opened up slightly for our application. Traditional, naturally aspirated engines usually require a 0.018-0.020–inch (top) and 0.016-0.018–inch (second) endgap. However, to prevent ring butting during the high heat experienced during nitrous use, the rings were endgapped at 0.026 inch (top) and 0.020 inch (second).

9 Because nitrous oxide generates added heat in the cylinders, proper block-cooling is important. The Summit cylinder block has been fitted with special top-end water restrictors (*arrows*) that help to better cool the cylinder bores. Traditional blocks have much larger upper coolant-passage holes that allow the water to circulate too quickly past the cylinder bores.

10 The heavy-duty splayed four-bolt main caps provide added bottom-end strength and crankshaft retention, but they do require a special bolt torque. To further beef up bottom-end strength, ARP bolts and studs were used. The longer ARP center studs were torqued to 63 lb-ft, while the shorter ARP splayed outer bolts were torqued to 65 lb-ft.

11 After the crankshaft was installed, the crank endplay was checked. Traditional, naturally aspirated engines usually run 0.004-0.006–inch crank endplay; however, with a nitrous engine you should slightly increase the endplay to about 0.006-0.008 inch.

12 A TCI Rattler harmonic dampener was used to quell parts-breaking harmonics. The TCI Rattler is fully SFI-approved for racing, and it has a variety of timing spec increments on its outer edge that aid in accurately setting the ignition timing, camshaft phasing, valve adjustment, and so on.

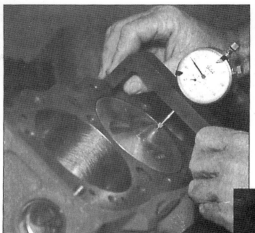

13 As with any performance engine, checking the piston deck height at top dead center (TDC) is important. However, because nitrous oxide use causes increased combustion temperatures, the top of the piston expands slightly more than usual. To allow proper clearance for this to occur, the piston deck height was set to be 0.006 inch below the block's deck height.

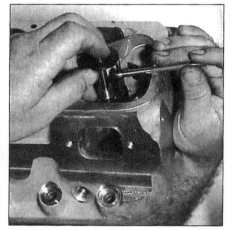

19 For outstanding airflow (needed for a large-cubic-inch, nitrous-breathing engine), a set of Air Flow Research aluminum cylinder heads (PN AFR210) was used. These AFR heads have big 210cc intake runners required to handle the flow requirements of this 383-cube nitrous engine. The fully machined heads were ordered as bare castings, then were fitted with a host of Competition Cams and Manley valvetrain gear that suited the needs of this nitrous engine. To ensure proper valve seal, special Teflon-lined seals from Sealed Power were used.

14 To generate big power, Porretta wanted to run a mechanical roller lifter camshaft. The pros at Competition Cams recommended a special nitrous roller cam (CS 3316-288-4 R12) was recommended for use with the nitrous engine. The cam specs are: 0.521-/0.550-inch lift (intake/exhaust), 236/244 degrees duration at 0.050-inch lift (intake/exhaust), and 112-degree lobe separation angle.

17 Camshafts "walk" slightly fore and aft during engine operation. When using nitrous, the cam walk is more aggressive as the engine instantly revs during nitrous use. To prevent excessive cam walk, a special camshaft end button was installed. Here the endplay between the cam and the Comp Cams aluminum timing cover is being checked. The endplay with this engine was 0.006 inch, which was within tolerance.

15 To ensure precise camshaft actuation, a Comp Cams geardrive set (PN 4100) was used. The geardrive set comes with various degree shims that allow for custom phasing of the camshaft.

18 Due to the fact that using nitrous oxide generates radical spikes in cylinder pressures, it is mandatory that top-quality head gaskets are used. With this 383 V-8, heavy-duty Fel-Pro head gaskets (PN 1038) were used.

20 Here the experts at McKenzie's Cylinder Heads (805/485-1810) check the installed height of the valve and retainer combo. The special Comp Cams valvesprings needed for the mechanical roller-lifter nitrous camshaft are extra burly, so checking for the proper installed height is important.

16 Using a professional-quality camshaft degree wheel, the camshaft was phased "straight up" in the block. If you don't have or don't know how to use a degree wheel, check out Comp Cams' complete cam installation kit (PN 4796), which includes the entire degree wheel kit and a VHS video that shows how to use the tools.

21 Todd McKenzie installed the heavy-duty valves, springs, retainers, and keepers. Afterward, the heads were flowed on McKenzie's Superflow bench, where the intake runners delivered 305-cfm airflow and the exhaust ports flowed 232-cfm airflow.

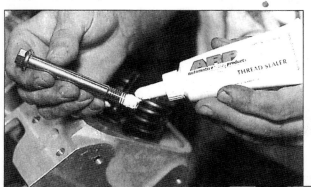

22 Due to the fact that nitrous oxide injection delivers extremely high cylinder pressures, heavy-duty ARP head bolts (PN 434-3601) were used. Be sure to apply thread sealer to the head bolts, because many of them go into the coolant passages of the heads. The ARP bolts (with thread sealant) should be torqued to 65 lb-ft.

Dyno Test

After building the 383 nitrous engine, we brought the combo to Duttweiller Performance (805/659-3648), where it was broken in, tuned, and then dyno-tested. The engine was first tested in a naturally aspirated format, followed by testing using the Nitrous Works nitrous system. As can be seen from the nitrous dyno figures, the nitrous system was activated at 4,000 rpm, where it made an immediate 241.6hp/298.1–lb-ft torque jump. It should be noted that the Nitrous Works nitrous system was jetted down to deliver about a 250hp hit. In actuality, the system can easily be jetted to deliver up to about 500 extra hp, but we chose to err on the conservative side. Note the headers used: They are special fenderwell headers from Hooker Industries for use with a '62–'67 Chevy Nova. All dyno testing was done using Borla XR1 performance mufflers to simulate real-world operating conditions. **CC**

23 An Air Flow Research stud girdle was used to help stabilize the valvetrain during high-rpm engine operation and when the extreme cylinder pressure spikes occur during nitrous use.

26 To ensure that plenty of air/fuel mixture would be delivered to the stroker V-8, a BG Fuel Systems Stage III 750-cfm double-pumper 4V carb (PN TM4779S3) was used. The BG carb is fully CNC-ported and features a host of hi-po upgrades such as high-flow throttle shafts, adjustable screw-in air bleeds, and special metering blocks.

	383 V-8		Nitrous 383 V-8	
RPM	HP	Torque	HP	Torque
3,400	285.9	441.6	281.0	434.2
3,600	304.1	443.7	311.7	455.2
3,800	321.3	444.1	326.2	447.1
4,000	340.9	447.6	567.8	**745.2**
4,200	365.5	456.9	578.1	723.0
4,400	392.0	467.9	603.2	720.0
4,600	413.2	471.7	625.2	713.8
4,800	431.5	**472.1**	637.1	697.2
5,000	446.6	469.1	651.2	684.1
5,200	463.5	468.2	659.4	666.0
5,400	473.7	460.7	667.0	648.8
5,600	479.0	449.2	**670.5**	628.8
5,800	477.5	432.4	665.9	603.1
6,000	**479.8**	420.0	655.4	573.7

24 As stated earlier, the increased cylinder pressure generated during nitrous use often generates added crankcase pressure. When extra pressure occurs, many times oil in the head's lifter valley will be forced out of the valve cover breather. To help deter this, small baffles were welded to the inside of the valve covers just below the breather port.

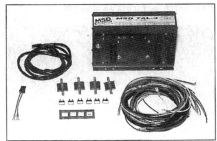

27 To prevent any high-rpm, belt-throwing mishaps, the engine was fitted with March Performance's (313/729-9070) trick serpentine-type beltdrive conversion kit (PNs 20130, 20140, and 6210). The kit features billet aluminum pulleys that are available in standard or underdrive ratio. Partially obstructed from sight is a Weiand Team G high-flow aluminum water pump.

Sources

Air Flow Research
Dept. CC
10490 Ilex Ave.
Pacoima, CA 91331
818/890-0616

BG Fuel Systems
Dept. CC
RR 1, Box 1900
Dahlonega, GA 30533
706/864-2206

Competition Cams
Dept. CC
3406 Democrat Rd.
Memphis, TN 38118
800/365-9145

Fel-Pro
(FP Performance)
Dept. CC
7450 N. McCormick Blvd.
Skokie, IL 60076
847/674-7700

MSD Ignition
Dept. CC
1490 Henry Brennan Dr.
El Paso, TX 79936
915/857-5200

The Nitrous Works
Dept. CC
RR 1, Box 1900
Dahlonega, GA 30533
706/864-7009

Summit Racing Equipment
Dept. CC
P.O. Box 909
Akron, OH 44309
800/230-3030

TCI (FP Performance)
Dept. CC
1 Equien Dr.
Ashland, MS 38603
601/224-8972

Weiand Industries
Dept. CC
2316 San Fernando Rd.
Los Angeles, CA 90065
213/225-4138

25 The most important aspect of this 383 nitrous engine buildup is, of course, the Nitrous Works nitrous system. The direct port nozzle system (PN 14020) uses two nitrous solenoids and two fuel solenoids to shuttle go-fast juice directly into the ports of a Weiand Team G intake manifold. This kit is adjustable between 150-500 hp, but we chose to jet the kit on the lower-horsepower side until the engine was fully tuned in. If plumbing your intake manifold for use with the port nitrous kit is more than you care to do, you can send your manifold to Nitrous Works where the work will be professionally done for an additional price.

28 Sparking the air/fuel mixture of a traditional street/strip V-8 is tough work, but igniting the high-pressure nitrous mixture is extremely difficult. Thus, this engine made use of an MSD 7AL-3 ignition system that not only provides a mega ignition zap but also incorporates a multistep timing retard needed for a serious nitrous engine. In addition, the 7AL-3 has a built-in three-step rev limiter along with multiple rpm switches.

Dragon Slayer

By Dave Emanuel

THIS SMALL-BLOCK CHEVY PUMPS OUT HORSEPOWER AND TORQUE LIKE A BIG-BLOCK—AND IDLES LIKE A STOCKER

Since its introduction in 1955, the small-block Chevrolet has been celebrated in word, picture, song, dyno test, and bank deposit. For over 30 years, it has reigned as the nation's most popular and prolific performance engine. And (moans and groans from the Ford and Chrysler contingents aside) the small-block Chevrolet has been the mainstay of the performance industry.

Much of the small-block's longevity is due to the fact that the engine is like a chameleon—it can adapt to any environment. From its original 265-cubic-inch displacement, Chevy's small-block has been worked and reworked, revamped and refined to fit the needs of the moment. During the years when bigger was equated with better, displacement grew to 283, 327, 350, and finally 400 cubic inches. When the need for downsizing or specific displacements arose, bore or stroke dimensions were squeezed back down, resulting in engine sizes of 262, 267, 302, 305, and 307 cubic inches.

In many cases, new displacements were achieved through the crossbreeding of previously existing parts. A 3-inch-stroke 283 crank installed in a block with a 4-inch bore results in 302 cubic inches. A 3¼-inch-stroke 327 crank stuffed into a 283 block (3.875-inch bore) produces 307 cubic inches. And a 350 crank (3.48-inch stroke) mated with a 3.736-inch bore adds up to 305 cubic inches. These combinations have all been used in production engines, but one of the most intriguing results of parts cross-breeding exists strictly within the province of hot rodding. With main bearing journals machined to a 2.45-inch diameter (from their original 2.65-inch diameter) the 3¾-inch-stroke crankshaft from a 400-cubic-inch small-block will drop right into a 4.030-inch-bore cylinder block and yield 383 cubic inches.

This combination makes for a killer high-performance street engine because the 3¾-inch stroke puts a mother-in-law-size bulge in the low-speed and midrange sections of the torque curve—at precisely the rpm levels where a street engine spends most of

Built by Nalley and Moody Racing Engines, this 383 small-block proved to be a dynamite street engine. In fact, with numbers such as 452 lbs.-ft. of torque, 431 horsepower, and 16 inches of vacuum at idle, it would make a few big-blocks turn green with envy. We won't mention the effect it has on the Ford and Mopar contingents. B&B polished aluminum rocker covers and chrome wing nuts topped off the Brodix cylinder heads, while a Bo Laws two-piece aluminum timing cover buttoned up the front of the engine. Cyclone Corvette headers with 1⅝-inch tubes routed exhaust gases out of the engine.

Illustration: Joe Goebel

Although a single-plane intake manifold and 850-cfm carburetor would have boosted top-end power, they also would have diminished low-speed torque. Since this engine was going to be used in a daily driven Trans Am, low-speed and midrange torque were of primary concern. Consequently, all tests were run with an Edelbrock Performer intake and Holley 750-cfm 4-barrel.

Manley forged Pro Series flat-top pistons fill the bores of a sonic-checked four-bolt 350 block. These pistons have the necessary compression height for a 3¾-inch stroke and 5.7-inch connecting rod. Fel Pro Print-O-Seal head gaskets were used to assure optimal sealing between the iron block and aluminum heads.

Small-block Chevrolets have almost bulletproof oiling systems, but oil can get trapped in the valley if the drain holes are restricted. A little work with a grinder is all that it takes to assure adequate drainage back to the pan.

its life. But torque does not live by stroke alone; power also rises and falls according to the choices made in camshaft, cylinder heads, intake manifold, and carburetor.

The standard approach to building a 383 is to start with a solid short-block and favor it with a healthy hydraulic cam and a set of reworked production cast-iron heads filled with 2.02-inch intake and 1.60-inch exhaust valves. Without question, such components provide impressive results, but the standard combination is just plain vanilla. With recent improvements in technology, it's possible to build a naturally aspirated engine that produces as much power as a turbocharged 350—at considerably less expense. By relying on a hydraulic roller camshaft, aluminum heads, and race-quality machine work and assembly, it's possible to achieve a turbocharger-caliber power curve with nothing more than a single 4-barrel carb. That's the theory, and after an extensive dyno session, that also turns out to be the fact.

Nalley and Moody Racing Engines of Norcross, Georgia, demonstrated the point with a 383 project that was done under the direction of Al Moody. The short-block was based around a 350 block with sonic-checked cylinder walls and four-bolt mains. Moody added a modified 400 crankshaft (with main journal diameter machined down from 2.65 inches to 2.45 inches) that had been meticulously custom-finished by John O's Automotive Machine in Douglasville, Georgia.

One of the shortcomings of a stock 400-cid small-block is its short 5.565-inch connecting-rod length. In high-performance applications, this rod places excessively high side loadings against the cylinder walls. To alleviate this condition, Nalley and Moody reconditioned a set of standard-length (5.7-inch), Chevrolet high-performance, forged-steel connecting rods.

You don't have to be a calculator wizard to figure out that a 5.7-inch connecting rod creates somewhat of a problem when combined with a crankshaft that swings its arms through a 3¾-inch stroke. Standard 350 pistons are obviously unusable because with their 1.560-inch compression distance (the measurement from the piston pin centerline to the piston deck) they would poke right through the cylinder heads at top dead center. But 5.7-inch rods are used so frequently in 383 small-blocks that Manley has released forged, flat-top "ready to run" Pro Series pistons specifically for this application. Manley part No. 49453 is designed to fit a 4.030-inch bore, and its 1.425-inch compression height allows it to be hooked to the longer rods and a 3¾-inch-stroke crankshaft. These pistons feature full floating pins, two valve reliefs, and 1/16-1/16-3/16-inch ring grooves, making them ideal for high-performance street use. Manley's matching ring package, which includes a plasma moly top ring, was also installed.

Other short-block components include Manley connecting rod and main bearings, a Manley high-volume, anticavitation oil pump, and a high-strength oil pump drive rod. Due to in-car space limitations, the block was bottomed off with a Chevrolet Z/28 four-quart oil pan and windage tray.

Like many engine builders, Nalley and Moody have found that it's not necessary to run high-viscosity oil in a high-performance street engine. They've found that there's some power to be gained by running 10W-30 racing oil, as opposed to 20W-50. Moody explains, "With a true racing engine [compared to a street engine] power output is a lot higher, and the engine runs at maximum power for most of its life. That makes a big difference to the bearings, and sometimes you just feel a little better with a thick oil because it gives the bearings a little more cushion. But it also costs some horsepower. We've found that an oil's additive package is the key to protecting an engine. It's really much more important than weight. We've used Valvoline's 10W-30 racing oil in a bunch of engines and have been real satisfied with the results. With lighter oil, you can also run tighter clearances for better oil control and increased longevity."

Another race-oriented component used on the short-block was a Fluidampr viscous vibration damper. Although this might be considered overkill for a street engine, the reason for its use is entirely valid. According to Moody, "Stock vibration dampers are tuned to the mass of the reciprocating assembly. A damper from a 400 won't do the job because the 350 pistons and rods don't weigh the same as the ones used in a 400. A 350 damper won't hack it either because it's designed for an internally balanced en-

Minimal deck clearance is one of the keys to producing impressive torque and horsepower. Although "zero deck" (piston level with the block deck) is the ideal, compression ratio consideration required a .005-inch depth in this case.

Many an engine misses its mark because of incorrect cam phasing, and all too frequently the timing chain and sprocket set are the cause. Cheap, off-brand sprockets may retard the cam up to eight degrees, and the chains in these sets stretch like a rubber biscuit. To avoid such problems, Nalley and Moody use Crane roller chain sets that are extremely accurate and offer proven durability.

Along with a hydraulic roller cam, the Dragonslayer 383 was fitted with Crane roller-tipped, variable-ratio rocker arms. As might be expected, maximum power output was achieved with the rocker ratio set to 1.6:1.

gine and doesn't have the counterweight you need for a 400 crankshaft. I know that 383s have been built with stock dampers, but I just can't see taking a chance on a damper you know is wrong. A lot of guys don't pay attention to the vibration damper, and they pay for it down the road."

Since the engine was ultimately to be installed in a legitimate "drive it to work every day" Trans Am equipped with an automatic transmission, valve timing had to be kept conservative. Crane offers several hydraulic roller profiles for both early and late small-block Chevy engines ('87 and earlier blocks require one type of cam, '88 and later blocks, designed for original-equipment hydraulic roller camshafts, require another), and Nalley and Moody have found that grind number HIR-278-2S-12 offers an excellent combination of low-speed driveability and top-end power.

Roller camshafts (especially hydraulic rollers) are particularly attractive in street applications because they can move the valve at much higher velocities than standard hydraulic tappet cams. This translates into broader power curves and a relatively smooth idle combined with super strong top-end power. There is also a power advantage gained from reduced friction—which is the basis for Detroit's recent switch to hydraulic roller camshafts. Moody noted that he's modified late-model engines for a number of his customers and they were amazed at the power increase brought on by a change in camshafts. Moody points out, "Some people think they're getting a lot of performance from the stock cam because it's a roller. They're really surprised when they find out that most factory hydraulic roller cams have valve timing that's identical to the standard hydraulic cams they replaced. There's really no great benefit with a factory roller cam; any power increase you get [compared to the same engine with a standard flat-tappet hydraulic cam] is strictly the result of reduced friction."

Performance aftermarket hydraulic roller grinds are dramatically different cams. They offer considerably quicker valve opening and closing rates than their flat-tappet counterparts, and therefore place more area under the lift curve. In practical terms, this means that a hydraulic roller cam opens the valves faster, lifts them higher, holds them open longer, and closes them more quickly than a conventional hydraulic lifter cam of the same duration.

Mated to the HIR-278-2S-12 cam was Crane's mandatory Cam-Ponent Kit, which includes hydraulic roller tappets, special-length pushrods, valve springs, steel retainers, machined steel valve-stem locks, valve-stem seals, a fuel pump pushrod, a bronze distributor gear, a cam sprocket bolt locking plate, and a needle bearing cam button spacer. Crane's roller timing chain and sprocket set was also included.

With only 222 degrees of intake duration at .050-inch tappet lift, the 278 camshaft is relatively short and therefore compatible with a stock torque converter. Consequently, compression ratio had to remain conservative to avoid excessive cylinder pressure and detonation. Although the heat-conducting characteristics of aluminum cylinder heads typically allow a ½- to full-point higher compression ratio than cast-iron heads, the wavering octane of pump gasoline dictates that some amount of safety margin be included. Based on their experience, Nalley and Moody felt that a compression ratio of no more than 10:1 was in order.

Which brings us to the cylinder heads. Although his company offers an aluminum street head, Brodix's J.V. Brotherton felt it was inadequate for a fire-breathing 383. As an alternative, he advised using a Brodix -8 head. This head was originally designed for oval track racing, but high-output street small-blocks have become so popular that Brodix also manufactures -8 heads with heat-riser passages for street use. And if they are not ported extensively, -8 heads—with 2.02/1.60-inch intake and exhaust valves—provide the high-velocity flow characteristics that are ideal for a street engine. With the -8 heads' combustion chambers worked to yield 73 cc's, and Fel-Pro number 1010 head gaskets (designed specifically for aluminum heads), compression was 10:1.

As soon as the 383 was fired on Nalley and Moody's SuperFlow 901 dyno, it was clear that something strange and wonderful was going on inside the test cell. Throttle response was unbelievable. Each time the dyno throttle lever was pushed forward, the 383 rapped like a race engine. And after it was broken in, it pumped out power numbers like a race engine, too. With an Edelbrock Performer intake manifold, a Holley 750-cfm carburetor, and an MSD distributor and 7AL ignition module in place, the 383 cranked out 432 horsepower at 5500 rpm and 452 lbs.-ft. of torque at 4000 rpm. Those are big-block numbers.

As impressive as the horsepower and torque figures are, they take on new significance when examined in the context of the total power curve. Most small-blocks that produce over 400 horsepower have power curves that rise sharply, hit the peak, then fall off quickly. And the peak is usually at 6500 or 7000 rpm. But this engine's horsepower curve not only peaked early at 5500 rpm, but it also stayed fat with over 400 ponies from 5000 to 5750 rpm. The torque curve was equally fat and sassy. It exceeded 410 lbs.-ft. at 2750 rpm and never dropped below that level up through 5500 rpm. What's more, it produced in excess of 440 lbs.-

DRAGON SLAYER

Combined with the hydraulic roller cam, Brodix -8 aluminum heads proved to be a killer combination. Although originally designed for racing, -8 heads are available with heat crossover passages for street use. With just a bit of cleanup work in the ports, these heads provide ideal port volume and velocity for high-performance street engines.

Crane dual valvesprings, steel retainers, and machined-steel valve locks keep Manley 2.02-inch intake and 1.60-inch exhaust valves in place. Other heavy-duty hardware include Crane 7/16-inch rocker studs and stepped pushrod guideplates.

Test Results
383-cubic-inch small-block Chevy tested on SuperFlow 901 dyno in acceleration mode at 200 rpm/second.

RPM	Corrected Torque	Corrected HP
2000	344	131
2250	368	157
2500	375	178
2750	411	215
3000	423	241
3250	437	270
3500	445	296
3750	451	322
4000	452*	344
4250	450	365
4500	446	382
4750	432	391
5000	429	409
5250	423	423
5500	411	431*
5750	372	407

*Peak readings.

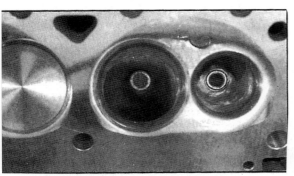

Brodix supplied the heads with the ports and combustion chambers completely finished. To maintain high port velocities, a minimal amount of material was removed. Combustion chamber volume was set to 73 cc's.

Before a 400 crankshaft can be stuffed in a 350 block, the main journals must be turned down from 2.65 inches to 2.45 inches. Of course, it never hurts to radius the oil holes as well.

Edelbrock's dual-plane Performer intake manifold contributed to the 383's strong low-speed and midrange torque. Power did drop off at 5750 rpm, but that was by design. Nalley's and Moody's approach to building street engines is to concentrate on torque first and horsepower second.

ft. from 3500 to 4500 rpm.

But equally impressive is the manifold vacuum level. Typically, when a small-block, even one displacing 383 cubic inches, is coaxed into producing over 400 horsepower, cam duration is usually so wild that idle quality deteriorates and manifold vacuum is so weak that power brake and vacuum accessory operation is erratic. With only 278 degrees of advertised duration, Crane's HIR-278 cam is relatively mild and pulled almost as much manifold vacuum as a stock cam. At 700 rpm, the 383 pulled 16 in./Hg; at 800 rpm, the reading was 17 in./Hg.

While the short cam duration pumped the low-speed and midrange power numbers up as if the engine had been given a shot of automotive steroids, it also limited top-end power. The dual-plane Performer intake manifold was another top-end-limiting factor, but both limitations were by design. A longer-duration cam and single-plane manifold would undoubtedly have bumped peak horsepower to over 450, but that increase would have come at the expense of power at lower engine speeds. Peak torque would have also taken a hit. Since this is a legitimate street engine, Nalley's and Moody's plan was to build as much torque as possible and simply let horsepower take its own course. But by combining the HIR-278 hydraulic roller cam with Brodix -8 aluminum heads, they achieved an extremely wide power curve, so they wound up with an engine that produced a ton of torque *and* a ton of horsepower—all at very livable rpm levels.

SOURCES

Brodix Cylinder Heads
Dept. HR
301 Maple
Mena, AR 71953
(501) 394-1075

Crane Cams, Inc.
Dept. HR
530 Fentress Blvd.
Daytona Beach, FL 32114
(904) 258-6174

Fel-Pro, Inc.
Dept. HR
7450 N. McCormick Blvd.
Skokie, IL 60076-8103
(312) 674-7700

Fluidampr
Dept. HR
537 E. Delavan Ave.
Buffalo, NY 14211
(716) 895-8000

John O's Automotive Machine
Dept. HR
8858 Bright Star Rd.
Douglasville, GA 30134
(404) 942-7212

Manley Performance Products
Dept. HR
1960 Swarthmore Ave.
Lakewood, NJ 08701
(201) 905-3366

MSD Ignitions
Dept. HR
1490 Henry Brennan Dr.
El Paso, TX 79936
(915) 857-5200

Nalley and Moody Racing Engines
Dept. HR
345 Lively Ave.
Norcross, GA 30071
(404) 368-0873

Here's some more food for thought. This was only the first step up the high-tech ladder. Nalley and Moody plan some future development work with Chevrolet's Tuned Port Injection (TPI) manifolds. Considering the positive effect that the long TPI runners have on torque, 500-lbs.-ft. small-blocks will be the wave of the future. **HR**

400 CUBIC INCHES FOR F-BODIES

INCHING UP ON PERFORMANCE WITH A 400-CID SMALL-BLOCK

By Rick Voegelin

When you get right down to it, there is really nothing wrong with a late-model Firebird or Camaro that an infusion of horsepower won't cure. GM engineers hit a home run with the styling and handling of the third-generation F-bodies, but there are some serious deficiencies in the engine compartment. The limited availability of High Output L69 305's and the de-tuning of port fuel-injected LB9's have produced a giant void in the performance of the F-bodies. But what are the alternatives for F-body owners who are saddled with lackluster, low-performance V8's? GM is bringing on T.P.I. 350's for '87, but if you can't wait or want to improve your existing car, how about taking a hot rodder's approach?

That's exactly what Firebird fanatic Herb Adams did. He came up with a fix that follows the first commandment of undercover performance: "Look stock and carry a long arm." In this instance, the long arm resides in a 400-cid Chevrolet short-block.

It takes a keen eye to detect the switch. To all outward appearances, the new engine is just another stock 305. It's the same subtle approach that made Q-ships the scourge of German submariners. And the subterfuge still works today, as anyone who's been snookered by an oversize small-block at the Wednesday night grudge races can confirm.

Regular readers of HOT ROD may have the feeling they've seen this particular F-body before. In a sense, they have. Adams was a participant in this magazine's second "Suspension Shootout" reported in the July and August '82 issues. By the time the bullets stopped flying, Herb had already made the decision to add a new Firebird to his fleet of F-bodies. This car is the twin to that original Shootout machine. Outfitted with a full complement of Herb's Very Special Equipment suspension components, it had the cornering power to rip up chunks of asphalt. But since it also had a standard LG-4 small-block (factory rated at 145 hp), it lacked the horsepower to even spin the rear tires on glare ice.

As part of the preparation for the original Shootout, Herb had upgraded his F-body's small-block with an assortment of aftermarket items. Bolt-on modifications included a dual-plane Holley aluminum intake manifold, a 600-cfm Holley 4-barrel with vacuum-operated secondaries, and a free-flowing exhaust system. These changes chopped 2 seconds off the Firebird's 0-to-60 acceleration times. Herb later installed a cam from an LU-5 fuel-injected engine to keep pace with the High Output motors being released by the factory.

After three years of magazine road tests and hundreds of circuits around a skidpad, the 305 was clearly showing its age. And although Herb found his reworked F-body to be an enjoyable automobile, he still hankered for the effortless acceleration he recalled from the large displacement supercars of the Sixties. A 400-cid small-block seemed the ideal solution.

The notion of building a 400-cid F-body is not new. Chevrolet engineers considered exactly that combination for the second-generation Z28, and even bolted together a running prototype in 1971. Large-displacement Pontiac and Oldsmobile engines were offered in various Super Duty and Trans Am Firebirds. But since the arrival of the third generation of Camaros and Firebirds, five liters has been the upper limit for F-body engine displacements.

Adams' own F-body fitness program centered on a new 400-cid short-block. GM calls this "partial engine assembly No. 1400888," and gets $755 for it at retail. Herb planned to minimize both his expenses and his car's down-time by transferring the cylinder heads, induc-

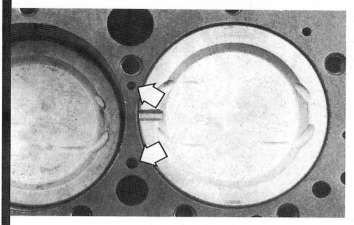

The 400-cid small-block has huge 4.125-inch diameter cylinders and a 3.75-inch stroke nodular iron crankshaft. Adjacent cylinders are "siamesed," so no water circulates between them. Extra steam holes (arrows) in deck surfaces prevent formation of air pockets. Cast pistons have .185-inch deep dish.

Adams' crew uses a 400-cid small-block head gasket to locate steam holes in late-model cylinder heads.

Matching steam holes must be drilled in late-model 305 cylinder head before installation on 400-cid short-block. LG-4 heads have nominal 58.9cc combustion chambers and 1.84/1.50-inch valves.

Chevy 400-cid small-blocks are externally balanced, so they require a counterweighted balancer and flywheel.

SOURCE
Herb Adams VSE
100 Calle Del Oaks
Del Rey Oaks, CA 93940
(408) 899-4859

PERFORMANCE SUMMARY

Modifications	0 to 60 mph (seconds)	quarter-mile (e.t./speed)
Stock LG-4	9.3	
Installed Holley 600-cfm carb, Dominator manifold, and cold-air induction	8.7	
Installed 3.0-inch high-flow single exhaust	7.3	
Installed 400-cid short-block with LG-4 heads and LU-5 cam	5.7	14.13/100

tion, and accessories from the 305 to the healthy 400 block.

The enduring beauty of the small-block Chevy is its parts interchangeability. Through 30 years of production, this amazing engine's external dimensions have remained unchanged. Bolting a 400-cid small-block into a late-model Firebird or Camaro is simply a matter of turning wrenches. Motor mounts, bellhousing, and exhaust manifolds are all a straight-across swap. Since the 400-cid small-block is externally balanced, it requires a counterweighted harmonic balancer (part No. 6272225) and a special flywheel (part No. 340298). These two items added another $100 to Herb's parts bill. An 11-inch clutch was installed out of respect for the 400's considerable torque output; this required relieving the inside of the bellhousing with a die grinder to provide clearance for the clutch cover. An early-model oil pan (part No. 3985999) is also necessary, since the 400 block has its dipstick on the driver's side. (Herb fabricated a new tube and gauge for the dipstick to eliminate interference with his Firebird's cast-iron exhaust manifolds.) Finally, Adams scrounged a starter that fits the 400's new flywheel.

Adapting the LG-4's cylinder heads to the 400 block requires drilling steam holes between the combustion chambers. Herb's crew used a 400 head gasket as a template; since the placement of the holes is not critical, a drill press is all the machinery required. (An enterprising hot rodder who had salvaged a complete 400-cid small-block could skip this step. In addition, cylinder heads installed on 400-cid small-blocks in the early Seventies have larger valves and ports than the late-model 305 castings, and thus offer greater performance potential.) Thanks to the 400's deep-dished pistons, compression remains a reasonable 9.5:1.

Driving the Firebird with its new powerplant was a revelation. With acceleration to match its sleek aerodynamics, the car became a latter-day supercar. Corners that used to require downshifts could now be negotiated with the help of the 400's irresistible torque. Passing became a pure delight, with no noise and no fuss, just set-you-back-in-your-seat acceleration.

These subjective impressions were confirmed at the test track. Zero-to-60 times dropped another 1.5 seconds, and the quarter-mile passed in a quick 14.13 seconds in the heat of a 100-degree afternoon. That's a full 2-second improvement in elapsed time over the car's box-stock performance. In deference to the 400's small valves, modest camshaft timing, and restrictive factory exhaust manifolds, the gear changes were made at only 4800 rpm! The effect on real-world fuel economy is hardly noticeable, since less throttle opening is required to produce the same power as the overworked 305. Finally, the swap is virtually invisible to prying eyes.

So why is Adams letting the world know his secret? Simple. The car has gone to a new owner. Consider yourself forewarned not to trust any "stock" small-block, no matter how original its corporate blue paint may appear. **HR**

BRUTE STRENGTH

How To Build A Street-Wise 383 Small-Block V8 That Delivers 450 Lbs-Ft Of Torque

By John Kiewicz

If you go to the local Friday night cruise-in, chances are that you'll hear the typical onslaught of bench racing—and the inflated engine-horsepower-output claims. One street machiner will claim to have 400 horsepower, while another person will brag about having 425. In reality, odds are that none of the braggers have ever seen an engine run on a dynamometer, much less tested their own engine. So, not only are the bench racers talking gibberish, they're bragging about the wrong *real* street power figure. Say what? We mean that if the bench racers were hip, they'd be talking about their engine's torque output, not the horsepower output. Why? Because on the street, hefty torque is what quickly launches your street machine away from the stoplight. Horsepower, however, is what keeps you going once you're at speed. With that said, let's turn to the focus of this story.

John Barkley, one of the ad-sales folk at CAR CRAFT, has a '72 Camaro that he built up for himself and his son. The car sports a 350-cube small-block Chevy engine, but it is lacking big power output after serving a hard life as a daily driver. Thus, Barkley decided to build a new 383 V8 powerplant that would make the Camaro a serious contender on the street. Being that the Camaro weighs in at a somewhat portly 3500 pounds and runs a mild 3.73:1 rear gear ratio, Barkley knew that the new engine should generate plenty of torque to deliver the best street performance. But, the real question was this: What combo of parts would not only deliver great torque but great driveability, as well? Rather than piece together a hodge-podge of parts, Barkley went with a proven parts combination from Edelbrock. Key components include Edelbrock Performer RPM aluminum cylinder heads, a cam and lifter set, an aluminum intake manifold and a 750-cfm carburetor. In addition, other top-notch parts (from companies including Childs & Albert, Crower, ROL and JE Pistons) were used to ensure that the engine would be bulletproof during high-rpm use. The engine was built by Larry's Performance in Montebello, California, which specializes in mail-order engines for street machines. Follow along as Roland Marquez at Larry's Performance builds John Barkley's new 383 torque monster V8.

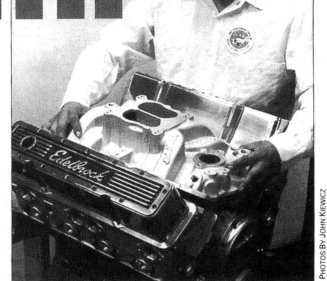

Roland Marquez at Larry's Performance installs an Edelbrock Performer RPM intake manifold (part of a complete Edelbrock parts package) onto John Barkley's new 383 V8. Delivering just over 450 lbs-ft torque, the 383 should make Barkley's '72 Camaro a serious contender on the street.

PHOTOS BY JOHN KIEWICZ

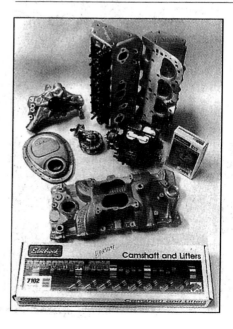

Most car crafters know that a good performing engine works well because it includes a matched set of parts. A big camshaft doesn't work well if the heads can't flow enough air, and if the intake and carb are too restrictive then the engine will be a dog. Thus, Barkley went with a power package from Edelbrock that includes Performer RPM aluminum cylinder heads (part No. 6073), a cam and lifter kit (part No. 7102), an Accu-Drive gear drive set (part No. 7890), a two-piece aluminum timing cover set (part No. 4242), a dual-plane aluminum intake manifold (part No. 7104), an 850-cfm Q-jet carburetor (part No. 1910), a hi-perf street fuel pump (part. No 1721) and a Victor series aluminum water pump (part No. 8812).

Bulletproof Bottom End

The 383 build up began with machining the cylinder block. Larry's Performance increased the 350 V8's bore from the stock 4.00 inches to 4.030 inches. Afterward, the bores were final honed using a torque plate that delivers more accurate results and better overall piston-ring sealing. The block was also fully deburred and painted with Glyptal, which aids in oil flowback.

To deliver 383 cubic inches, the stock 3.48-inch stroke crankshaft from the 350 V8 was replaced with a Larry's Performance cast-steel crankshaft that delivers a 3.75-inch stroke. Because the larger stroke places the connecting rods farther outboard, the cylinder block must be ground down to gain proper clearance. Not much grinding is needed, just enough to clear the outer rod bolt as it passes near the bottom edge of the cylinder bore.

Rather than reusing the stock main-cap bolts, the engine was upgraded using a Milodon engine stud kit (part No. 81117). Installing the Milodon studs is easy, but just remember to bottom tap all of the bolt holes first to ensure proper thread engagement.

Most Chevy 383 stroker engines use a 3.75-inch stroke crankshaft swiped from a 400 small-block V8. But, to make it fit properly in a 350 block, you need to machine down the 400's 2.65-inch main journals to 2.45 inches, which is the stock main-journal size of a 350 small-block crankshaft. However, Larry's Performance now offers a new 3.75-inch stroke crank that is a specifically cast unit; it incorporates the smaller 2.45-inch main journals, which are the standard diameter for a stock 350 block. With the Larry's Performance crank, simply order standard 350 V8 main bearings (such as part No. M-3500-STD from Childs & Albert), and the crank drops right into place without any fancy machine work.

After the main bearings are positioned in the block, carefully lower in the crankshaft. Notice the rubber caps placed over the ends of the Milodon main studs. The caps reduce the chance of gouging the crankshaft if it accidently comes into contact with the studs.

Once the crankshaft is in place, install the main caps. Thread the Milodon nuts onto the main studs, and then torque the main caps incrementally to the proper spec. Many rookie engine builders tighten the main-cap bolts/nuts beginning at one end of the engine and working downward. This is incorrect. Rather, torque the bolts/nuts in sequence, starting from the center and working outward in a radial pattern.

When the main caps are properly tightened in position, check the crankshaft endplay at a few different locations. To do so, carefully move the crankshaft rearward in the block as far as it will go, and use a feeler gauge between the crank and the thrust bearing surface to check the endplay. Then, move the crankshaft forward as far as it will go, and check the endplay again in various locations. Usually, for performance engines such as this, crank endplay should be around 0.005-0.008 inch.

Rather than reusing the stock 350 rods, Barkley upgraded to Crower Sportsman Stroker rods (part No. SSP91302) because of their much greater strength. The Crower rods *(left)* are forged from 4340 chrome-moly steel and come with heavy-duty 7/16-inch bolts. The Sportsman Stroker Rods feature an extra 0.250-inch cam-to-rod clearance over traditional Sportsman rods. In addition, Barkley opted for Crower 6-inch rods rather than traditional 5.7-inch-length rods *(right)*—the added length increases the piston's dwell time at TDC, which helps generate added power.

For optimum sealing, the 383 engine was fitted with Childs & Albert ZGS (Zero Gap Second) piston rings. Top ring endgap varies depending on the application (street, nitrous, super-charged and so on), so carefully measure and adjust the endgap as needed. Always measure the endgap with the ring positioned about 1 inch down the cylinder bore. After you have measured, adjusted (by grinding the end of the ring) and remeasured the gap, be sure to keep the ring paired with the same cylinder in which it was measured. Doing so ensures that the rings are custom-matched to each cylinder, thus generating added reliability and power.

After the ring endgaps have been set, carefully install the top ring and the ZGS second ring on the piston. The ZGS design uses an overlapping step-type ring end that greatly reduces blow-by. C&A says that when using ZGS rings, leak-down testing shows a reading of only about one percent. This ultralow amount of blowby generates about nine percent more horsepower and about 18 lbs-ft torque over-stock-type rings, according to C&A.

Once the rings are properly installed, carefully tap the pistons into the cylinder bore using a ring compressor and the wooden handle of a hammer. For this engine, JE forged aluminum pistons (part No. 132682) were installed that deliver a 10:1 compression ratio when used with a traditional 64cc cylinder head combustion chamber. The JE pistons feature an inverted dome that incorporates the proper amount of valve relief area to avoid piston-to-valve damage. With aftermarket pistons, the height of the piston pin differs, depending on the length of the connecting rod used. Thus, because this engine uses Crower 6-inch rods (in place of the stock 5.7-inch rods), these JE pistons feature a slightly higher pin height.

Oiling System

When building a high-performance engine, always upgrade to a high-volume oil pump. Because of increased bearing load, rpm and horsepower generated with a street/strip engine, the added flow, pressure and reliability of a high-volume oil pump is cheap insurance for a bulletproof engine. For this engine, a Milodon oil pump (part No. 18750) was used that provides a 25-percent increase in volume and pressure over stock oil pumps.

BRUTE STRENGTH

To improve oil control, this engine was fitted with a Milodon windage tray. The windage tray (part No. 32100) is a simple bolt-on that prevents oil from splashing against the crankshaft, thus affecting power output. Because the mounting height and design of windage trays vary, rotate the engine through its firing order while checking to make sure that the rods do not come into contact with the windage tray. Notice how this rod *(arrow)* comes close to hitting, but it does have about ¼-inch clearance, which is fine.

To cap off the oil system, the 383 engine received a new Milodon Low Profile oil pan. The oil pan (part No. 30900) features a triangulated sump design that has a 6-quart oil capacity, but it doesn't hang down and cause a ground clearance problem. Plus, the Milodon pan has internal baffles that control oil movement during hard launches and when the car leans under cornering.

Camshaft

Being that the cam plays such a major role in the power output of an engine, rather than guess at the proper camshaft, Barkley opted to run an Edelbrock Performer RPM cam. The Edelbrock cam kit, specifically developed through extensive dyno testing, was designed to deliver a specific power output when used with good-flowing cylinder heads, such as Edelbrock Performer RPM aluminum heads. Using traditional 1.5:1 rocker arms, the cam specs out as follows: 0.488-/0.510-inch lift (intake and exhaust, respectively) with a duration of 234-/244-degrees at 0.050-inch tappet lift. As always, before installation, be sure to give each of the camshaft lobes a generous slathering of cam lube (included in the Edelbrock kit) to prevent the cam from going flat upon engine break-in.

Installing an Edelbrock Accu-Drive camshaft gear drive is easy. Simply align the top-gear bolt holes so that the camshaft dowel pin passes through the relief in the gear. Notice the small offset bushing. In each gear-drive kit, Edelbrock includes a variety of different offset bushings that allow the camshaft to be adjusted in two-degree increments. To phase the camshaft differently, simply insert the proper bushing, and you're done.

Once the cam and crank timing gears are installed, position the main idler gear so that it fits between the upper and lower gears. If you notice, the idler gears are not the same size (the smaller gear should go toward the driver's-side of the engine). Besides delivering extremely accurate camshaft actuation, the Accu-Drive also eliminates the problems associated with a worn-out/stretched timing chain.

One of the neatest parts we've seen in a while is Edelbrock's two-piece stamped aluminum timing cover. With traditional timing covers, if you needed to change or service the timing gears and/or camshaft, you had to remove the oil pan before removing the timing cover. With Edelbrock's two-piece timing cover, install the special backing plate on the front of the block. Then, install the oil pan as you normally would.

Once the special backing plate is in position, bolt on the main timing cover using the supplied bolts. That's it, you're done. Although the timing cover looks stock from the outside, it can be easily removed *without* having to drop the oil pan.

Harmonics can lead to a quick death of a street/strip engine. Carefully balancing the engine parts (such as the crank, rods and pistons) is a noteworthy start, but you should really consider upgrading to a high-performance harmonic balancer. This 383 engine build-up included a new Fluidampr that contains a special internal gel that quells unwanted vibrations. Plus, the Fluidampr has a complete range of crankshaft degree markings on the outside that aid in tuning and degreeing parts.

Edelbrock Cylinder Heads

As CAR CRAFT has preached before, always use top-quality head gaskets with a street/strip engine build-up. To ensure reliable performance, ROL Pro-Torque head gaskets were used for this torque monster 383 V8. Although this engine will initially run naturally aspirated, down the road it may be upgraded with a nitrous oxide kit. Thus, installing the ROL head gaskets now is rock-solid insurance against gasket failure later on when the nitrous button is pushed.

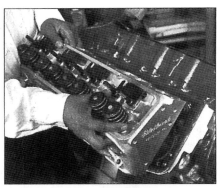

Once again, Barkley chose to install Edelbrock Performer RPM aluminum cylinder heads because they were part of a dyno-tested, tried-and-true power parts package. The heads use the latest intake and exhaust runner configurations specifically designed to perform well on the street. In addition, the Edelbrock heads are constructed of aluminum, which not only sheds about 45 pounds off the front of the engine, but also dissipates heat faster, thus reducing the chances of detonation.

After the Edelbrock heads are installed and torqued to spec, install the pushrods while paying keen attention to properly aligning the pushrod end within the center of the lifter. A dab of assembly lube on both tips of the pushrod will provide proper lubrication upon break-in. Notice the intake runners on the Edelbrock heads. When you order a set of Performer RPM heads, they come with CNC port-matched and hand-blended intake and exhaust runners that deliver outstanding flow.

BRUTE STRENGTH

Because the engine would be seeing high rpm regularly (along with an occasional hit of nitrous oxide), the engine was fitted with Crower stainless steel roller-rocker arms.

The Crower rockers not only reduce valvetrain friction with their fully rollerized fulcrum and roller tip, but they are also super rigid to thwart valvetrain deflection. Notice the screw-in studs and guideplates—they're standard with the Performer RPM heads.

Although single-plane intake manifolds flow great at high rpm, they're usually not the best choice for low-rpm daily driving. Thus, this 383 stroker engine made use of an Edelbrock Performer RPM dual-plane intake that delivers optimum low rpm power but still provides plenty of airflow to 6500 rpm.

To ensure ample spark delivery, the cruddy stock distributor was tossed in favor of a Carburetor Shop HEI distributor. Although GM HEI distributors are well noted for their ample spark output, the Carb Shop HEI has been fitted with a slew of secret upgrades. Remember that with higher compression and nitrous oxide use, you're going to need a healthy ignition system to properly ignite the air/fuel mixture.

What Makes A Good Gasket?

You've got too much invested in your engine to let heat and detonation rob you of engine life and performance. Thus, spending a few extra bucks on quality engine gaskets is a good investment to protect your street/strip engine for the long haul. ROL Torque Master gaskets now offers a graphite gasket specifically for small-block Chevy engines. Starting with a rugged steel core, ROL combines layers of flexible graphite with stainless steel firerings, and tops off its design with a silicone coating and Pozi-seal beading (pointing finger). According to ROL, "Graphite is the logical choice due to the superior heat resistance, up to 2100 degrees F, and graphite transfers heat across the entire gasket face, which eliminates hot spots and contributes to lower operating temperatures. Additionally, the stainless steel firerings will withstand 200-degrees-higher combustion temperatures than conventional combustion armor." A fundamental characteristic of this design is that the body materials are clinched to the core, not glued, and will accommodate thermal expansion and motion without the delamination and tearing that are common to some gaskets.

Because this engine will be used in a street car, Barkley chose not to use a big ol' honking race-type carb. Rather, he opted for the top performance and rock-solid reliability of an Edelbrock Q-jet carburetor. Delivering 850 cfm of airflow, the Edelbrock Performer RPM carb will be able to feed the hungry 383 cube small-block without a problem. Notice the Q-jet's electric choke and PCV ports. Touches like this are what make the Edelbrock carb a user-friendly part for street machines that serve as daily drivers.

Final Look

Finished off with Edelbrock Elite Series valve covers and Trans-Dapt aluminum pulleys, the 383 engine has the perfect all-business look. Hedman headers, coated with Jet-Hot coating, will do a great job of venting unwanted exhaust nasties when the engine screams to its 6000-rpm redline. To guarantee that the engine starts every time, a high-performance gear-reduction starter from IMI Performance Products was used in place of a heat-tempermental stock starter.

Dyno Test

RPM	Horsepower	Torque
2500	176.8	371.5
3000	221.2	387.2
3500	264.2	396.5
4000	344.6	452.4
4500	383.0	447.0
5000	411.6	432.3
5500	423.9	404.8
6000	414.0	362.4

Pump It Up

To ensure that the engine would have a steady supply of fuel during high-rpm operation, an Edelbrock Street Fuel Pump was installed. The Edelbrock pump incorporates a special valve design that improves flow quantity and quality, and it delivers a flow capacity of 110 gallons per hour—enough to support about 550 horsepower.

Because this 383 V8 will serve duty in a daily driver, keeping the engine cool will be of utmost importance. An Edelbrock Victor Series aluminum water pump not only sheds a few pounds off the nose of the car, but it also dissipates heat better. Plus, Edelbrock water pumps use a special, larger-diameter, CNC-machined impeller that delivers outstanding pressure and volume—even at low rpm. Within the pump housing, the internal passages feature rounded radii and equalized passages that deliver a higher coolant velocity.

Sources

Carburetor Shop
Dept. CC
8460 Red Oak St.
Rancho Cucamonga, CA 91730
909/481-5816

Childs & Albert
Dept. CC
24849 Anza Dr.
Valencia, CA 91355
805/295-1900

Crower Equipment Co.
Dept. CC
3333 Main St.
Chula Vista, CA 91911
619/422-1191

Edelbrock Corp.
Dept. CC
2700 California St.
Torrance, CA 90503
310/782-2900

Fluidampr/Vibratech Inc.
Dept. CC
11980 Walden Ave.
Alden, NY 14004
716/937-3603

IMI Performance Products
Dept. CC
10065 Greenleaf Ave.
Santa Fe Springs, CA 90503
310/944-9265

JE Pistons Inc.
Dept. CC
15312 Connector Ln.
Huntington Beach, CA 92649
714/898-9763

Larry's Performance Shop
Dept. CC
2008 Whittier Blvd.
Montebello, CA 90640
213/721-6284

Milodon Inc.
Dept. CC
20716 Plummer St.
Chatsworth, CA 91311
818/407-1211

ROL Manufacturing
Dept. CC
154 F St.
Perrysburg, OH 43551
800/233-6272

Trans-Dapt
Dept. CC
16410 Manning Way
Cerritos, CA 90701
310/921-0404

AT LARGE

By Tom Madigan

Building A 454-Cubic-Inch Small-Block

WHY BUILD A 454-CUBIC-INCH small-block when you can buy a good 454-cubic-inch big-block? It's a perfectly logical question, and the answer turns out to be as reasonable as any red-blooded American hot rodder could expect. It's one-upmanship! Engine builder Paul Pfaff of Huntington Beach, California, had a customer who wanted to one-up a competitor in a street rod association. Pfaff decided that the man's '34 Ford coupe needed something of the never-been-done-before variety, hence the decision to build a 454-cubic-inch small-block Chevy.

There are large-displacement aftermarket aluminum small-blocks as well as large-displacement small-blocks built with a spacer and sleeves to achieve the necessary volume, but to Pfaff's knowldege, there had not yet been a 454 built from a factory-stock cast-iron block.

No, it isn't done with mirrors. It's done with stroke length; 4⅛ inches of it to be precise. A Chevrolet Bow Tie block (PN366287) and a Sonny Bryant B.R.E. fully counterweighted 4330M alloy billet crankshaft are the heart of this combination. Bore holes were drilled in the crankshaft's counterweights, permitting the installation of Mallory (heavy metal) steel inserts and a standard full internal balance job. The Bow Tie block began

Pfaff's 550 hp 454-cubic-inch small-block has as much displacement as you're likely to see in a small-block Chevy. Using the latest cylinder head and EFI technology, it delivers big-block power from a small-block package.

Because of the high pin placement in the piston, a 0.030-inch steel spacer is added to support the oil ring. Pistons are installed at 0.005 inch clearance.

Side clearance is set at 0.016 inch. This view also shows how the notches ground in the oil pan rails provide extra clearance for the connecting rods.

Considerable machine work was required in the crankcase area to ensure adequate crankshaft clearance (see arrows). The block was notched and relieved along the oil pan rails and at the bottom of each cylinder adjacent to the camshaft for connecting rod clearance.

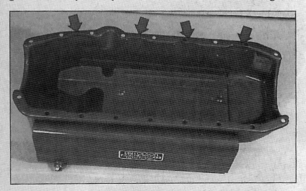

Note how the pan rails in the Dan Olson Racing Products oil pan are notched to match the clearance notches that are ground in the block.

Big cubes call for a big bore and a long stroke. The Bryant Racing Enterprises 4330M alloy crank offers a 4.125-inch stroke, full counterweighting, and the addition of Mallory metal slugs to facilitate internal balancing.

JE pistons yield a compression ratio of 10.5:1, which is livable on the street thanks to the wonders of electronic fuel management. High-buck 6-inch Carrillo rods were used in this engine, but cheaper aftermarket substitutes are available.

At Large

life as the 4⅛-inch bore variety and received a whopping 0.060-inch-overbore to achieve a bore size of 4.185 inches.

The enormous stroke involved made it necessary to seal the bottom portion of the water jacket so that notches could be machined in the lower portion of the crankcase to clear the crankshaft and connecting rods. After sonic checking the cylinder walls to verify their integrity with the large overbore, 2 inches of aluminum epoxy was poured into the bottom of the water jacket. No other changes were required since the block already features siamesed cylinders, blind head-bolt holes, and splayed four-bolt caps on the three center main saddles.

Required machine work to clear the crank and rods included special clearance notches along the pan rail as well as the portion of the cylinder wall adjacent to the camshaft. The rough, unfinished cylinders were bored using a centering plate fixture to correctly maintain the factory 4.400-inch bore spacing. Cylinders are finished with a 600-grit stone and a torque plate in place. The deck surface is refinished after special water hole plugs were installed to increase rigidity and regulate water flow.

JE dished pistons deliver a compression ratio of 10.5:1, and Speed-Pro plasma moly rings seal the cylinders. The connecting rods are 6-inch Carrillo units that feature a special "K" cut to clear the small-base circle Crane camshaft. The small-base circle of 0.860 inch provides a considerable amount of clearance when compared to the standard base circle diameter of 1.100 inch. Standard Vandervell bearings were chamfered for radii clearance and installed. Dan Olson created a special 10-quart oil pan with built-in notches to clear the rods.

The aluminum 356-T6 Phase VI Bow Tie heads were ported in-house at Pfaff Engines. Crane roller rockers and lifters were used with the roller lifters to incorporate an offset cup for guide clearance. Manley severe-duty stainless-steel valves were installed with 2.100-inch-diameter units on the intake side and 1.600-inch valves on the exhaust side.

The inlet system comprises an Edelbrock EFI manifold base with modified ports, a Lingenfelter box plenum with high-flow runners, an Arizona Speed & Marine throttle body (58mm flowing 1000 cfm), and an N.O.S. nitrous oxide injection system.

Pfaff estimated 550 horsepower from this engine, and he evaluated it with both the fuel-injected version and a baseline version that uses a single four-barrel carburetor. This particular combination is a high-dollar setup to be sure, but the instructive concept illustrates the

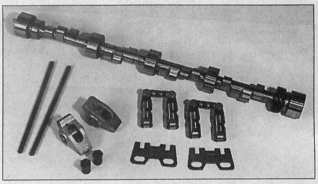

The Crane roller camshaft features 256/266-degrees duration for the intake and exhaust respectively at 0.050-inch lift. Advertised duration is 306/296 degrees. Net valve lift is 0.596 inch on the intake valve and 0.582 inch on the exhaust. Crane offset roller lifters and rocker arms are used to gain pushrod clearance with the ported heads, and Isky's adjustable pushrod guide plates keep everything aligned properly.

A giant 4.185-inch bore is evident in this view. Note the minimal deck area between cylinders and drilled water jacket plugs to control cooling. Huge intake ports on the Chevrolet Bow Tie Phase VI aluminum cylinder heads promise plenty of airflow to feed the beast.

Combustion chambers were mildly massaged and outfitted with Manley stainless-steel valves. The intake valves are 2.100 inch in diameter; the exhaust valves measure 1.600 inch.

These D-shaped exhaust ports also show a considerable amount of work by the Pfaff crew. Crane valvesprings and roller rockers control the valves.

type of modifications required to build a large-displacement small-block. Although you may never choose to duplicate this engine, you might encounter some of the clearance problems associated with stroker engines, and this buildup shows you how to handle them. Moreover, it also shows that the small-block is capable of delivering big-block power levels on a per-cubic-inch basis.

Pfaff prepares to put the big little block through its paces. The original goal was 500-550 hp with a streetable package.

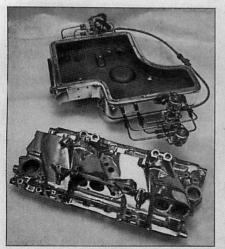

A polished Edelbrock TPI manifold base has been ported and fitted with a Lingenfelter box plenum and high-flow runners that have been plumbed for nitrous oxide injection. A 58mm Arizona Speed & Marine throttle body is also part of the package.

A K&N filter was attached to the throttle body for the initial dyno tests which were run with Pfaff's 1⅝-inch dyno headers.

Source
Paul Pfaff Racing Engines
Dept. CC
5362 System Drive
Huntington Beach, CA 92649
714/894-7573
FAX 714/894-7461

Test Results

The initial test on the 454 revolved around three different induction systems, each of which had its own strong points and weaknesses. For the guy who put all his money into the engine and has little left to spend on an induction system, the single four-barrel version is the least expensive choice. It makes good power and torque, particularly in the higher rpm range where the four-barrel intake manifold was able to move more air than the semi-siamesed runner manifold from Arizona Speed & Marine. While the 454 runs well and produces exceptional high-end power, it isn't meant to be a high-rpm engine. It all depends on your particular requirements.

The Arizona Speed & Marine unit on the other hand, clearly produced an abundance of torque and horsepower at the lower end of the rpm scale where it is often most useful. Moreover, it maintains this power advantage throughout most of the useable power band, falling behind the carbureted version only at the very upper end of the scale. Keep in mind that this is a lot of displacement to feed with a small-block induction system originally designed for engines in the 350-cubic-inch range.

The Lingenfelter box plenum manifold provided the best midrange and high-rpm results primarily because it has shorter runners to help high-end power and a large volume plenum to feed the monster. In a nutshell, the carbureted version is probably the least expensive, the Arizona Speed & Marine is the best overall for stump-pulling low-end torque, and the Lingenfelter unit is best for overall high-performance street use requiring a broad power band. Each intake system has its own particular characteristics, and you can easily see how they affect power at the rpm levels that interest you.

Holley Single Four-Barrel Carburetor

RPM	HP	Torque
2500	272.9	489.1
3000	292.3	511.7
3500	326.7	490.3
4000	427.1	560.8
4500	500.4	584.1
5000	534.7	561.8
5500	553.7	528.8
5900	549.8	489.2

Peak HP: 553.8 @ 5800 rpm, Peak Torque: 584.1 @ 4500 rpm, Average HP: 533.1, Average Torque: 424.6

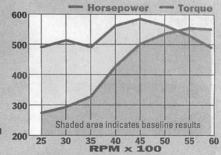

Arizona Speed & Marine Semi-Siamesed Runners

RPM	HP	Torque
2500	243	510.8
3000	317.4	555.5
3500	355	532.6
4000	435.5	571.7
4500	489.3	571.1
5000	502.8	528.2
5200	504.2	510.3

Peak HP: 502.8 @ 5000 rpm, Peak Torque: 582.9 @ 4200 rpm, Average HP: 457, Average Torque: 545

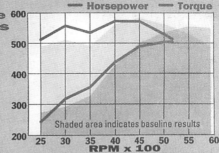

Lingenfelter Box Plenum Manifold

RPM	HP	Torque
2500	231.8	486.8
3000	311.8	545.8
3500	350.4	526.0
4000	429.4	563.7
4500	504.9	589.3
5000	527.2	553.9
5500	519.4	496.2
6000	483.2	425.3

Peak HP: 525.3 @ 4900 rpm, Peak Torque: 589.3 @ 4500, Average HP: 423.2, Average Torque: 536.7

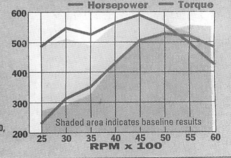

The Immaculate Deception

Cylinder Block

To build this engine you need a 4-inch bore large journal block in reasonably good condition. The standard 4.001-inch bore will yield 377 cubic inches with the 3.75-inch stroke of the 400 crankshaft. A 4.030-inch bore will yield a 383ci engine like ours, and a 4.060-inch bore will provide a whopping 388 cubic inches. The 400 crankshaft will fit into most 350 cylinder blocks after the large main journals have been turned down to the 350 journal size (see crank section). Some early two-bolt main blocks may interfere with the large counterweights, but minor grinding will cure the problem. The best block to use is a well-seasoned four-bolt main block, or lacking one of these, a new high performance four-bolt block from Chevrolet (PN 3970016). We used the special high tin alloy block (PN 366246), but its expense may be prohibitive for most street machiners. It is also a "green" casting that will make less power until subjected to considerable run-in. You must use a 350 or 327 block of the large journal variety because the 400 crank will not accept the large cut required to turn it down to small journal size. Hence you should avoid any blocks made prior to 1968.

Once you have settled on a final displacement, the block should be subjected to the standard preparation procedures. It should be deburred, thoroughly cleaned, bored, honed and align honed if necessary. The decks should be cut parallel to the crank centerline, but care should be taken that the cut is the absolute minimum required to achieve parallel decks. The more you cut here, the more you have to take off the top of the pistons later. Our new block required all of these operations, which were capably handled by the crew at Valley Head Service in Northridge, California. With all the special operations required for this engine, everyone in the shop was involved with it at one

High performance cylinder blocks are already notched for connecting rod clearance. Once the main journals have been ground to the correct size, the 400 crankshaft drops in with no counterweight or connecting rod interference. The standard length connecting rods (5.7 inches) will hit the camshaft on some cylinders and minor grinding is required.

Standard two-bolt and four-bolt blocks are not always notched and may require grinding to provide clearance for the connecting rod bolts. A small notch is all that's necessary, but the final determination must be made by dummy-assembling the engine for clearance checking.

CC's Underground Camaro Gets a Strongarm Powerplant for Street Supremacy

By John Baechtel

In these overregulated times it's often difficult to identify a truly contemporary approach to hot street performance. Turbocharging probably gets the nod, but street supercharging continues to gain in popularity even as more and more street machiners look to modifications that stress fuel economy. The problem facing most street machiners is the continuing lack of an established middle ground. Those cars built with a generous application of economy pieces are reasonably snappy performers and save gas, but very few of them are capable of standing up to a well thought out turbo setup or even an early high-compression, 4-barrel supercar. Nitrous oxide is one of the best ways to put the snarl back into these economy-oriented street machines, but even generous doses of it aren't enough to help overcome a serious competitor. Late-model cars are worse off than most since they may have to contend with yearly inspections that require the presence of all OEM smog equipment in a functioning capacity. Headers, intake manifolds, carburetors and other equipment are all part of the high performance plan, but the sight of them is often grounds for unwarranted hassling by inspectors.

So what can we do to regain some of that old zip while maintaining some sort of quasi-legal facade that will keep the bad guys off our case? Furthermore, can we do it with an engine that could pass a smog test if necessary? We thought it might be interesting to find out, so we undertook the construction of a super street engine designed to fulfill these requirements. There are a number of ways to approach the problem, but we selected one that has become increasingly popular—a large displacement, long stroke small-block based on factory parts. Paramount in our consideration was a need for gobs of torque and a high reliability factor. And since our 1980 Z/28 test car was to be the recipient of this killer piece, it seemed reasonable to duplicate the original engine's external appearance as closely as possible. So return with us now to those glorious days of yesteryear when sleepers were in style and everyone had a 283 under the hood—if you know what we mean.

Real old-timers will tell you that the best path to performance is to "put an arm in it," or as we say, use the longest stroke you can get away with to ensure lots of torque. Throw in another tradition in the form of lots of cubic inches and you have the makings of a serious contender. Now, you can build a 400ci small-block with standard length rods and it will be a hard runner. You can build a 350-based motor with a 400 crank and the short 400 rods and it will also yield good performance. In fact, you can build about 10 different combinations involving these large displacement small-blocks just by jockeying blocks, stroke length, rod length and piston compression height. We happened on the 383ci engine by accident, but it turned us on and that's what we built. The combination is basically a 400 crank in a 350 block with standard length rods and a reduced piston compression height. We've seen them before and we know they run very well. And we have a sneaking suspicion that the ultimate street weapon may indeed be a long arm and a can of corporate blue engine paint.

---

point or another. Owner Larry Ofria was quick to stress the importance of checking the dimensional integrity of the new casting, and some of the problems we encountered reinforce this concern. The main saddles were off by a significant amount and precision realignment was called for before the decks could be trued. And even after the block was rendered dimensionally correct and the crank ground to the proper specs, it would not fit into the bearings and was not about to turn. A bit of examination was obviously in order.

It was determined that tolerance in the block casting was such that standard grinding and radiusing of the main bearing journals still did not provide enough clearance in some locations. The crank cheeks rubbed ever so slightly on the block bulkheads. The crank was put back in the grinder and the throw surfaces and radii were reground just enough to avoid the interference. Now the crank fit in the bearings and turned relatively easily, but a tight spot was felt and required further investigation. Finally the problem was traced to rear main cap distortion, evident when the cap was torqued to factory specs. When torque was backed off 10 ft.-lbs., the tight spot went away and the crank turned freely. But a very slight thump could still be heard as the crank was revolved. After much head scratching, frowning and mutual "I dunno's," we uncovered the problem—one that had not been encountered on previous

The arrow indicates the shiny spot where the outer surface of the thrust flange was contacting the inside of the rear maincap. Since the 350 maincap is smaller than the 400 maincap, the fit is relatively close and bears checking.

engines of this nature. A very small bump on the perimeter of the thrust flange was rubbing on the inside of the rear main cap. It was the last possible place you would expect a problem, but you have to remember that we were putting a 400 crank into a 350 block. The 400 block has larger main caps because of the larger 2.65-inch main journals. The thrust flange clears easily in a 400 main cap, but it gets very close in a 350 main cap. In most cases they clear, but the tolerance stack involving the tight cap and a slightly irregular thrust flange was enough to trip us up. Watch for it on your engine.

Crankshaft

The crankshaft is the heart of this powerplant. It is the standard cast crank from a 400ci small-block. There is nothing special about it other than the fact that it offers a 3.75-inch stroke which is highly attractive in terms of torque production and stout street

In the configuration we have assembled, the engine could be internally balanced, but it is easier to use the standard 400 balancer which provides external balancing. The 400 flexplate or flywheel should also be used.

The Immaculate Deception

performance. The crank is available over the counter under PN 3951527, and most engine shops have good ones in stock if your cash flow situation requires a used item. What you are looking for is a crank that is reasonably straight with minimal rod

Automatic cars will make use of the standard GM flexplate for 400ci engines, but stock equipped cars should pass on the factory flywheel and use a high performance, explosion-resistant flywheel like this one from McLeod Industries Inc. The part number is 460231 and it is designed for an externally balanced application like the 454 or 400. Make certain you have it in time for your machinist to balance it properly with the rest of the internal engine parts; balancing is imperative with one of these engines.

journal wear. Main journal wear or damage isn't particularly significant in light of the large amount of material being removed. Once you uncover a likely prospect, take the time to have it Magnaflux inspected to ensure the absence of cracks. This is particularly important since you're going to be reducing the amount of overlap between the journals when the mains are reground. This doesn't appear to significantly weaken the crank, but a crack in the wrong spot may be enough to do you in if it isn't caught.

Other than main journal preparation, the crank is subjected to standard crank prep operations including deburring, oil hole chamfering, micro polishing and thrust clearance during mock assembly. It is also wise to have the flywheel flange checked to ensure that it is perpendicular to the crank centerline. The crank requires no other preparation other than balancing after the rest of the internal pieces have been modified and their weights determined.

Oiling System

The lubrication system is configured to match the car. Since the Z/28 is something of a handler, we anticipate more than a few curves, so we went with Moroso's road racing oil pan which is at home in either a left or right hand turn—or going straight. The screen windage tray was used in lieu of a factory unit since it tends to grab the oil and keep it from bouncing back into the crank assembly. A 1¾-inch extension and corresponding pump drive were used to lower the pump to the bottom of the sump. Their bottom cover pickup (PN 2490) was added to the stock Z/28 oil pump to eliminate a separate pickup tube.

Bearing clearances have been

Moroso's screen windage tray required the use of two extra maincap bolts with extended studs. Note the 1¾-inch oil pump extension and the bottom cover pickup installed on the stock oil pump. The square cutouts in the screen windage tray are for connecting rod clearance. Here again, Specialty Fasteners studs are used to secure the oil pan.

kept relatively tight at .002-inch on the rods and .0028-inch on the mains. A .030-inch hole is drilled in the front center oil gallery plug to provide additional timing chain lubrication, and the oiling holes in the main bearings were drilled to match the size of the openings in the cylinder block.

Connecting Rods

The rods we selected for this engine are the standard Z/28, large journal rods (PN 3973386). You can use whatever rod you like as long as it is a large journal unit configured for a pressed pin application. The rods we used are those commonly referred to as "pink rods" because of the pink paint applied at the factory. They are the best choice because they receive additional preparation and Magnafluxing and are hand selected for the least amount of flash on the rod beams. Valley Head Service removed the stock bolts and installed high-strength bolts and nuts from Specialty Fasteners. Their new rod bolts have been tested to very high tensile strength and are recommended for this type of application. We did not polish the rod beams, but we did have to grind four of the rods where they hit the camshaft. This occurs on cylinders 1, 2, 5 and 6 and requires slight material removal from the rod near the bolt head. The only way to determine this accurately is during mock assembly of the engine. If you're using a small base circle cam like a mild roller profile, grinding may not be necessary, but stock cams and large lobe rollers are going to hit the rods. And don't take our word for it; check each and every rod and cam lobe by rotating the engine slowly with the cam installed and properly phased to the crankshaft. It all goes together very smoothly, but you have to check it. For final assembly we used TRW CL-77 bearings fitted at .002-inch.

Pistons

We've already said that the key to this engine is the long stroke and large displacement, but here is another item that helps bring it all together. Any 350 piston will bolt right into this motor if you use the short 400 connecting rods, which incidentally clear the camshaft and make the whole thing a bolt-together operation. We wanted to use the long rods (standard

These photos show the amount of material removed from the top of the factory pistons to make them fit the engine. Photo (A) is a comparison between the original domed piston and the flat top version prior to valve notch cutting. A side view, (B), shows how the dome and part of the deck surface are removed. There is still plenty of material left in the deck surface, and deep valve notches are still possible. Notice how this modification automatically moves the ring placement higher on the piston.

5.7-inch) for several reasons. First of all, we wanted to avoid the severe rod angularity incurred with the short rods. We also wanted to spin the motor fairly fast if necessary, and the Z/28 rod is a significantly better piece than the 400 rod and better rod bolts are available.

But when you stack up the long stroke with a long rod and a standard 350 piston, you end up with a piston that sticks out the top

The Immaculate Deception

of the block by about .100-inch. Now, normally a factory piston is about .030-inch down the bore at TDC, and when you add these figures together you come amazingly close to the difference in rod length between the standard 5.7-inch rod and the 5.565-inch 400 rod. Obviously we're going to have to whack off a bunch of deck material if we're ever going to put cylinder heads on the engine. The trick is in determining just how much material to remove and which pistons can safely handle the loss without collapsing under fire.

Ideally we want a minimum of .035-inch piston-to-cylinder head clearance, but we're also planning on using Chevrolet's recently released "O-ring" type head gasket (PN 14011041) which compresses to .040-inch. So in our case we can shoot for a zero deck height and minimize the amount of material removed from the piston deck. We used the factory LT-1 pistons because we wanted to maintain a factory parts approach as much as we could, but there are a lot of other pistons that will do the job. The factory piston and TRW's PN L2304F probably come up the lightest because they have more material removed. If you wanted to pick up some additional deck thickness for added reliability, you could use TRW's 3 9/16-inch stroker piston for the 350 (PN L2278F). It features a higher pin placement which lets you take less material off the deck surface. This piston and their Super Stock piston, PN L2327F, are both racing models and may require slightly more piston-to-wall clearance than some of the others. We haven't checked these other numbers, but some of them may provide a less expensive alternative since they are powerforged replacement pistons. If your dealer permits, try comparing the PN L2304F piston with the following numbers: PN L2417F, PN L2403F, PN L2369F and PN L2256F. In a Speed Pro piston you might investigate PN 2363P, PN 2363PA and PN 2244P.

In any case, a portion of the top of the piston has to come off and you only determine the amount by trial assembly and actual measurement of each piston as it protrudes from the block. Write it on the top of the piston and make certain you have them numbered correctly and that you match the pistons to the rods each time.

After the pistons are cut, you'll have to go back and determine piston-to-valve clearance. We determined that a .100-inch deeper valve notch was not unreasonable with the factory piston and Valley Head Service cut and blended the notches for us. When you do this it is important to measure the angle of the valve in the cylinder head and transfer this angle to the valve notches in the piston. By matching these angles, the piston will contact the valve squarely should valve float occur, and you may escape with little or no damage, depending on the severity of the valve float. Make certain you blend and smooth all the sharp edges on the piston top and then have the decks glass-beaded to obtain a uniform surface. Install them in the motor with .0035-.004-inch piston-to-wall clearance.

Use TRW's single moly ring set (PN T7945M) with the top ring gap set at .018-inch and the second ring set at .012-inch. This ring set has standard tension oil rings which require no special fitting. Be certain the rings are installed with the small dots facing up. Spray the cylinder walls with 3-In-1 oil, install the rings dry and lightly coat the piston skirts with engine oil.

Camshaft & Valvetrain

Chevrolet engineers felt that we should try one of their factory grinds in this engine. We weren't opposed to doing so based on the stout performance turned in by a factory cam in our 327ci bracket motor earlier this year. They suggested the first-design off-road cam (PN 3927140) which we have used successfully in 327ci and 350ci

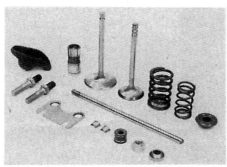

Chevrolet's high performance valvetrain hardware is more than adequate for an all-out street assault. Swirl polished valves, dual valvesprings, hardened retainers and pushrods, screw-in studs and guideplates make up a reliable yet formidable package.

engines in the past. They felt that the 140 cam would be very smooth and tractable in the large displacement, torque-efficient engine. The corresponding valve lifters are the piddle valve variety (PN 5232695). Z/28 pushrods (PN 3796243), and Chevrolet's dual valvespring and hardened retainer, PN 330585 and PN 330586 respectively, are used to control the valves. Aluminum retainers are generally regarded as something of a no-no for street performance applications, but you can get away with it with the Chevy pieces

The factory camshaft was installed four degrees advanced with the aid of a Cloyes true roller timing set that features three separate keyways for timing adjustments. Note the front cover studs from Specialty Fasteners.

because the retainers are hardened and there is no valvespring dampener to chew up the aluminum. When installing factory guideplates, it is a good idea to lightly dress and round the inside of the notches on a small belt sander. They are stamped pieces that sometimes have rough surfaces that can damage even the hardened pushrods. Moreover, you want to avoid any binding they might cause. Make certain each pushrod spins freely in the guideplate.

Cylinder Heads

Although a reasonably strong engine could be built with the early 1.94-inch valve, straight plug cylinder heads, we opted for the large valve 292 turbo castings because of their improved port configuration. For our purposes, Valley Head Service suggested that their "Streetmaster Porting" would provide the best overall performance. In what they refer to as their "Power Play" approach to engine building and head porting, they felt that the cylinder heads would provide most of the significant gains for our engine.

On the intake side (A) they work into the port approximately 1½ inches, blending and opening the port to match the intake gasket (Mr. Gasket PN R-102). The valve pocket is reworked slightly from the base of the valve guide to the valve seat, but the port is not ported all the way through because they want to maintain a rough surface for improved atomization. Furthermore, there is no need to drastically enlarge the port since this would kill the velocity required for good street performance.

A

The exhaust port **(B)**, on the other hand, is ported all the way through because it needs all the help it can get. It is given a smooth surface, with the short side radius being very critical. The port floor is made as wide, flat and high as possible to optimize exhaust flow. The combustion chamber is polished and the area on the intake side of the spark is laid back slightly.

Their "Marathon" valve job **(C)** is a three-angle approach with .060-inch intake seats and .090-inch exhaust seats, both beginning .015-inch from the edge of the valve. Final cc'ing and milling brings the chambers to 64cc's. The Chevrolet dual valvesprings were assembled to within ±.005-inch of the installed height required for proper spring pressure (135 pounds at 1.720 inches).

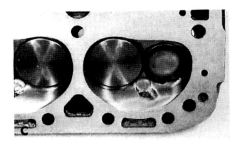

Compression Ratio

Since flat top pistons give the most desirable flame travel, we weren't particularly distraught about chopping off the piston dome and part of the deck, but we had to juggle head gasket thickness, deck height and combustion chamber volume to arrive at a desired compression ratio. With premium gas and fuel additives, 11.0:1 doesn't seem entirely unreasonable, and our CR works out to slightly more than this with the zero deck height, .040-inch thick compressed gasket thickness and a 64cc combustion chamber. This also takes into account the deepened valve

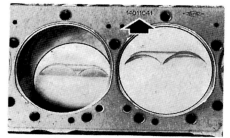

Chevrolet's special O-ring head gasket is designed for use with 4 and 4⅛-inch bores. It compresses to .040-inch thickness and the extra area helps reduce compression on our 4-inch bore engine. The 2.05-inch intake valves and 1.6-inch exhaust valves required larger valve reliefs than the original valve clearance slot in the factory pistons.

notches. The combination of overbore, reduced deck height and small chamber volume has placed us back in the vicinity of the 11.0:1 CR originally obtainable with the factory piston. If you want less compression, use a larger combustion chamber. You can also add a little deck height, but don't get crazy here, because you've already taken a healthy slice off the top of the piston. The turbo heads we used normally come with approximately 71cc chambers that would get your CR down to a reasonable figure if necessary. Compute it on paper first and you'll be able to determine the exact procedure to follow to achieve the CR you want.

Intake Manifold

We all felt that an aftermarket manifold would be necessary to feed this monster, but it didn't really fit into our deceptive plan. That's why we're using Chevrolet's aluminum Quadrajet manifold with a few subtle modifications. When this machine rolls into an inspection station resplendent in corporate blue engine paint, with a stock looking Quadrajet carb and an EGR valve on the manifold, it's going to take a well-trained eye to spot the ruse. There are still some obvious clues, but we're not talking. Marvin Miller plumbed the hidden nitrous system under our manifold so there's no plate to expose us. The fittings go in through the back of the block where they are easily hidden beneath the ignition system. Gene Kuklovski at Marvin Miller informed us that the modification is much easier to perform on the aluminum manifold and that owners of cast iron manifolds can expect to pay a good deal more for the same plumbing. We elected not to port-match the manifold at first because we wanted to see how well it worked with the heads in a stock condition. The ports are quite a bit smaller than the ports in the heads,

Marvin Miller provided these special bulkhead fittings for their nitrous oxide injection system. The block was extra thick in the area where we chose to install them. We thinned it with a hand grinder, drilled two ⅜-inch pilot holes, and spot-faced the surface with a screw-in stud cutter. The holes were then drilled to ½-inch and the fittings installed with silicon sealer. The lines clear the number eight pushrods and the distributor housing with room to spare.

A stock aluminum Quadrajet intake manifold was delivered to Marvin Miller for transformation into a direct port nitrous injection setup. They found it necessary to cut away part of the exhaust heat crossover passage and weld in a flat plate to seal it. The manifold was then plumbed with nitrous and fuel lines.

A nitrous line and a fuel line are run to each port, but the ports get different nozzles according to their entry angle. Runners leading from the lower chamber of the intake plenum enter the port at a flatter angle; thus they receive extended nozzles that are drilled to direct the extra fuel and nitrous oxide downstream. The steep runners from the upper chamber of the plenum have nozzles that are already aimed slightly downstream via their entry angle into the port.

Underground Camaro

Corporate blue engine paint (PN 1052442) is an important part of our deceptive plan, but it is often difficult to find. Many GM dealers don't care to stock it because of government shipping regulations regarding paint and other materials. A very close substitute is Krylon's GM Blue (PN 1930). Be certain to paint the entire engine, and if you really want to get sneaky, try installing a set of Targetmaster decals on the valve covers.

of the engine's real capability if fitted with serious induction and exhaust system hardware. But there's really no need for it. We're already pushing 400 horsepower with the stock intake manifold and well over 400 ft.-lbs. of torque in the mid-range. The nitrous oxide injection system will easily put the engine into the grey area where parts breakage becomes frequent. Once the engine is in the car, we have a feeling it's going to show us that the cylinder heads are actually a little too big and the cam is a little too wild. They work well with a good manifold and headers, but we think the car is going to want a smaller cam and smaller heads and probably a torque converter to help overcome the 3.42:1 gear ratio. In fact, one of Kenne-Bell's "Switch Pitch" torque converters seems like just the ticket for hot street performance with this combination. We're getting into a lot of maybes here, but it's really the only way to pin this thing down. At any rate it's certainly better for us to do all the legwork and make all the mistakes so that you only have to do it once—the right way. It's all coming up in following installments on CC's Underground Camaro.

and the mismatch is pretty substantial. Still, we're actually dumping into a larger runner after the manifold, and the mismatch may help slightly in the reversion department, which we know to be a significant problem with this manifold. If dyno time is available, we may port-match the manifold and try a high-velocity aftermarket manifold to observe the difference.

For What It's Worth...

If you study the parameters by which our underground powerplant was built, it quickly becomes evident that the entire effort is a massive compromise. No one in their right mind would go with a stock intake manifold and carburetor on top of an engine like this one. And even more objectionable is the fact that we're going to run through the stock exhaust manifolds and the stock exhaust system, catalytic converter and all. The compression ratio is way too high for unleaded fuel, but we only have to worry about that when we baby the car in for a landing at the emissions test center where everything has to look just right. We can't really recommend the use of dummy converters with straight-through pipes, or a dual converter setup, but we're willing to concede that a set of headers and a similar exhaust system arrangement would be highly beneficial.

Power output is certainly acceptable with the configuration we're using, and the addition of an Edelbrock Streetmaster manifold provided an increase indicative

383ci Chevy

Factory Parts Guide

PART NUMBER	ITEM AND/OR ASSEMBLY
366246	Cylinder block, 4-bolt mains, tin alloy
3951527	Crankshaft, 400ci engine, 3.750-inch stroke
3973386	Connecting rods, Z/28, large journal
3942543	LT-1 piston, .030-inch oversize
3965784	Cylinder heads, bare 292 turbo castings
14011041	Head gasket, integral O-ring, .040-inch compressed thickness
3796243	Pushrods, Z/28 type with hardened tip
3973418	Guideplates
3973416	Rocker arm stud, screw-in type
3764389	Valve guide seal
330585	Dual valvespring for HD camshafts
330586	Aluminum retainer (hardened)
3947770	Keeper, purple
361976	Intake valve, 2.05-inch diameter
3849818	Exhaust valve, 1.6-inch diameter
6272225	Balancer for 400ci engine
340298	Flexplate for 400ci engine, auto trans
1103291	Corvette distributor, HEI unit
14007377	Aluminum Q-jet intake manifold
8629423	M-40 Turbo 400 trans, Z/28 type
8626648	Torque converter, high performance

Dyno Results

Stock Intake Manifold

RPM	HP	Torque
3000	254	444
3500	283	424
4000	320	421
4500	372	434
5000	389	409
5500	387	370
6000	385	337

Edelbrock Streetmaster Manifold

RPM	HP	Torque
3000	246	431
3500	284	426
4000	326	428
4500	382	446
5000	401	421
5500	408	389
6000	403	353

Specialty Equipment Sources

Autotronic Controls Corp.
6908 Commerce
El Paso, TX 79915

Cloyes Gear and Products
4520 Beidler Rd.
Willoughby, OH 44094

Marvin Miller Manufacturing
7745 S. Greenleaf Ave.
Whittier, CA 90602

McLeod Industries
1125 N. Armando St.
Anaheim, CA 92806

Moroso Performance Products Inc.
Carter Drive
Guilford, CT 06437

Specialty Fasteners
2601 Redlands Dr.
Costa Mesa, CA 92627

Valley Head Service
19340 Londelius St.
Northridge, CA 91324

454-inch Small-Block CHEVY

The small-block takes on new status by growing as large as 482 cubic inches

TEXT AND PHOTOS BY DON GREEN ■ You'll find very few people who will argue with the statement that the small-block Chevy is the world's most popular performance engine. After being around in the same basic form for more than 17 years, most of the bugs have been worked out. Not that there were all that many bugs to begin with. Light, compact and capable of producing bunches of horsepower in relation to its displacement and weight, the small-block Chevy has, at one time or another, found itself in virtually every type of racing vehicle from Indianapolis to Irwindale, Budd's Creek to Bonneville.

If anyone has ever had a complaint about the engine, it was generally that it was "too small." It simply couldn't be built into the large displacements that some people felt were necessary for particular forms of racing. The introduction of the 400-cubic-inch version of the small-block in 1970 helped quiet some of the critics, but development of the 400 has been slow due to the lack of factory-produced, performance-oriented engine parts (such as forged-steel cranks and high-compression pistons). But even at 400 cubic inches, there are people who insist that the engine still isn't big enough or capable of being strong enough.

There's one guy around, however, to whom that thought hasn't even occurred. The only thing he knows is that small-block Chevys are the way to go, and if you work at it long enough, you're going to find a way to make them as big as you want. He's currently building and running small-blocks that range in displacement from 427 to 482 cubic inches, the most popular being a 454-cubic-inch version.

The man's name is Jack Conely, owner and operator of Conely Speed

Hand-ground smooth radius helps prevent main web cracks, and notches are for rod clearance.

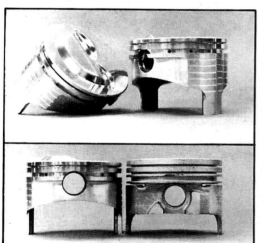

Special domed J.E. forged pistons (top) have only two ring grooves, are domed for high compression. Pin location (above) is higher than stock.

Walls of intake ports are milled away, replaced with welded plates to straighten and improve flow.

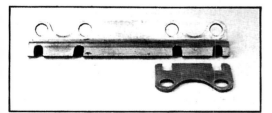

Special pushrod guide plates are required to help reposition rods after moving intake port walls.

Shop, and a Chevy man all the way. His shop is located in Brighton, Michigan, about a half-hour's drive northwest of Detroit and, oddly enough, quite close to the General Motors Proving Grounds.

By choice, Jack is a roundy-round racer... and a good one. He builds and drives his own cars, as well as building engines for himself and many other East Coast and Midwestern racers. He is a one-man racing effort whose personal racing experiences have carried him as far as the Indianapolis 500 in an attempt to compete against the heavily sponsored, multi-million-dollar teams. But Conely's real love is racing his Super Modified, one of those narrow, tubular, open-wheeled cars with a minimum of sheet metal covering and an unlimited engine. It was for the Super Modifieds that Conely got into his extensive small-block development program. The result, a rash of broken lap records wherever the car is run—even against competition like Offys, Fords and big-block Chevys. No one can believe the power produced by the small-block engine, and few are willing to believe that Conely's engines could possibly have as large a displacement as the rumors claim. But they do.

Our first knowledge of Conely came from a Chevrolet engineer who mentioned the 454-inch small-blocks. That was impressive enough until Conely himself told us that he had built a pair of 482-inchers for a customer. But 454's are all Jack personally needs to dominate the Super Modified circuit in the northeastern United States. The engines are completely reliable and produce more than enough power to win consistently, much to the embarrassment of many big-block Chevy owners. Now, with the revision of drag racing's Pro Stock rules to include the compact cars and small-block engines, Jack is looking forward to lending some of his ability and experience to drag racing. Current plans call for some dyno work on carbureted small-block combinations suitable for Pro Stock and Modified Production classes to back up his knowledge of Super Modified-type injected motors that are ideally suited to Gas and Altered classes.

The engines that Jack builds for himself have a 454-cubic-inch displacement and, like all the large displacement small-blocks he builds, are based on the relatively new 400-cubic-inch cylinder block. For the 454-inch displacement, the stock 4-1/8 (4.125) inch bore is enlarged to 4-3/16 (4.1875) inches, an increase that has proven very acceptable in racing engines. For the larger engines such as the 482-cubic-inch

454 SMALL-BLOCK

version, the bore must be enlarged to 4.250 inches. Conely warns, however, that only *good* castings will accept a 4.250-inch bore. He suggests checking the cylinder walls of the casting with a dial indicator to determine the amount of core shift. If there is more than a .030-inch shift, the .125-inch overbore should not be attempted. If the shift is under .030 inch, the boring bar should be adjusted to compensate. As a rule of thumb, a properly done 4.250-inch-bore block will have a finished cylinder wall thickness of .110 inch.

Jack handcuts a 3/16-1/4-inch radius on the main webs at the point where the webs meet the cylinder banks (see photo). This radius removes the stock three-angle cut that the factory uses that often becomes the starting point for main web cracks. The sides of the main caps are milled to make visual detection of cracks easier during assembly and maintenance of the engine. The increased crankshaft throw also makes it necessary to grind notches at the bottom of the cylinder bores for rod clearance. Conely considers it an "absolute must" to O-ring the large-bore cylinder blocks.

As you might imagine, the crankshaft is a very non-stock part. Conely uses Moldex cranks exclusively, each one being completely machined from a billet of certified 4340 aircraft-quality steel, and each costing about $600. The 454's have a finished stroke of 4-1/8 inches, while the 482's go out to a full 4-1/4 inches. The Moldex cranks are completely crossdrilled and tufftrided, and have a 1/8-inch radius ground into the crank pins. No center counterweights are used. Balancing is accomplished by the installation of pieces of extremely heavy and expensive Mallory metal, since the cheaper and more common method of welding weight to the counterweights only adds to the crankshaft's clearance problems. The mains are ground to 2.648 inches; the throws are ground to 2.199 inches.

It's only fitting that a 454 should use 454 connecting rods, so Jack begins with selected, high-performance 454 big-block connecting rods. These forged-steel big-block rods are over one-half inch longer (center-to-center) than stock 400-inch small-block rods, and .435-inch longer than high-performance 302-350 rods. The long crankshaft stroke requires that a lot of material be removed from the big end of the rod in the area of the bolt holes for camshaft clearance. The rod cap is

Heat-treated billet crank for small-block 454 is shorter, 10 pounds lighter than big-block crank.

Heavy slugs of Mallory metal are used in counterweights for balancing the long-stroke crankshafts

Pushrod holes are cut out to allow for relocated pushrods; slots also help oil drainback slightly.

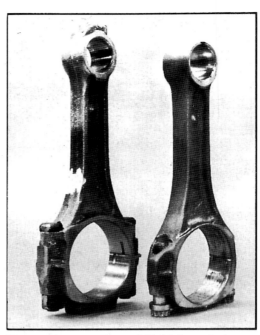

Race-ready connecting rod (right) is made from a selected, stock, 427-454 big-block rod (left).

Connecting-rod cap bosses are machined down for clearance, grooved for added strength (right).

Stock boss on small end of rod (left) must be cut off for inside piston clearance, oil hole added.

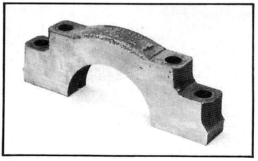

Milling side surfaces of main bearing caps makes visual detection of cracks easier, can save engine.

also reshaped around the bolt holes for cylinder-block clearance. The reinforcing pad across the bottom of the rod cap is machined down slightly for reduction of weight and improved clearance, and a radius is milled across the pad for increased rigidity (see photo). The stock rod bolts are replaced with special 7/16-inch, 4340 steel bolts. The forging marks are removed from the sides of the rods, and the small ends are radiused for weight reduction and interior piston clearance. An oil hole is also drilled in the rod's small end for improved lubrication to the floating piston pin. Finally, the entire rod is shot-peened for fatigue resistance.

Since there aren't any stock pistons available for the bore/stroke combinations Conely uses, he relies on custom forgings supplied by J E Pistons in Monterey Park, California. The combination of the long stroke and long rods places the pin hole unusually high in the forging, but this is beneficial in controlling the piston as it moves up and down in the cylinder. Only two piston rings are used: a single 1/16-inch dykes compression ring; and a 1/8-inch oil ring. The high pin height requires that the oil ring pass across the pin hole. Locating proper rings was also something of a problem, so Jack machines his own from moly castings supplied by TRW (chrome rings are used for better abrasion resistance in engines that are run on dirt tracks). The full-floating piston pins are special tapered items supplied by J E in a .990-inch diameter and a stock 427 big-block Chevy length. The pins are held in place by .072-inch-thick Spirolok retainers. With the gasolines now available, Jack usually settles for a maximum true compression ratio of 11.5:1.

To handle the large displacements, considerable work is done to the small-block cylinder heads. Dyno testing has shown an increase of 15 horsepower for the angle plug heads compared to heads with straight spark plug holes (both heads having the same chamber volume).

Conely's shop performs quite a bit of work on the ports beyond the usual porting and polishing. A mill is used to cut away the walls of the intake ports near the intake manifold surface (see photo). Iron plates are then welded into the head castings to replace the intake port walls, but the new plates are positioned so that the ports are effectively made larger and straighter. The new walls of the ports are moved enough so that they obstruct the stock intake pushrod guide holes, therefore the guide holes are slotted so that the pushrods can be moved slightly. The slotting also helps oil drainback to the

454 SMALL-BLOCK

cam valley. Offset buttons are also required in the tops of the intake valve lifters to help compensate for the pushrods, as are special offset guide plates which Conely manufactures. The whole thing becomes very obvious when the complete head is seen on an engine with the rocker arms in place and the valve cover

Offset lifter buttons (left) help locate pushrods moved during intake port work.

The amount the pushrods are offset is obvious when rocker arms are installed.

removed. The exhaust rocker arms remain in their straight up/down position, while the intake rockers are canted at an angle (see photo).

Surprisingly, Jack uses smaller valves than most people would suspect in an engine of the size he runs. The intake valves range between 1-15/16 and 2 inches, smaller than stock 302 valves. The exhausts have a diameter of 1-11/16 inches. Conely has tried just about any valve combination you can name, including the tuliped MoPar hemi-style valves, and has settled for the above combination. He admits that the intake and exhaust valve pockets in the pistons and knowing how to unshroud the valve seats have a lot to do with making the small-block breathe. In any installation using the 400 cylinder block, steam vent holes *must* be drilled in the heads to match the stock vent holes in the block between the siamesed cylinders.

There are very few people in the country who have done as much development on the small-block Chevy cylinder head as Conely. Besides all the unique valve combinations he has tried, he was working with open-chamber small-block heads as far back as ten years ago, and has experimented with canted-*valve* small-block heads.

Every Chevy enthusiast has heard of the legendary aluminum small-block heads—heard of them, but never seen any. Jack runs a set of them on his Super Modified engine. The best part is that the set he runs isn't

Finished combustion chamber shows some unshrouding and relative valve sizes.

his only set—he has lots of them, and they can be bought from him if you really want them or need them badly enough. Current plans call for limited availability of aluminum canted-valve (not just canted-plug, canted-*valve*) heads produced from his stock of inline valve aluminum heads. The canted-valve heads produce a horsepower increase above 4000 rpm

Jack's camshaft selections are based on his circle-track background. He uses Crower cams exclusively, most of them being roller

tappet grinds. His favorite is a special 648-R grind cam (not listed in Crower's catalogue) that they grind for his large-displacement engines. It's a dual-pattern cam having a 320-degree intake duration and a .328-degree exhaust duration. The intake and exhaust lifts are .610 and .620 inch, respectively. Jack also uses Crower's roller lifters, retainers, pushrods and rev-kits.

Small-block windage tray must be modified to clear the ZL-1/427 big-block oil pump.

Early model steel crankshaft gear is way to go; later cast-iron gears break easily.

The valve springs are special items Conely has produced from chrome-vanadium wire. The springs are a three-piece design (outer, inner, damper) and feature a .020-inch interference fit. The spring ends are also shaped to prevent them from gouging the retainers and possibly releasing aluminum chips into the oil.

Continued on page 35

Street Rod SPECIAL

QUADRADEUCE MOTORVATION

BUILDING AN ALL-ALUMINUM, 520HP SMALL-BLOCK FOR A RADICAL STREET ROD

By Will Handzel

Hot rodders are a creative bunch. They are driven by the urge to build and race something that has never existed before just because no one else has done it.

An excellent example engendered by this kind of thinking is the Summit Racing Equipment QuadraDeuce—a very bitchin' highboy roadster that has an all-aluminum small-block driving all four wheels through a six-speed transmis-

The aluminum block (part No. 1013440) is from Chevrolet, as are the valve covers (part No. 10185064), but the aluminum cylinder heads are TFS Twisted Wedge pieces that have been reworked for the heavy breathing, 400ci engine. Dyno figures with the TWM/DFI fuel injection, MSD ignition and Hooker headers showed 520 hp at 6000 rpm and 480 lbs-ft of torque at 4500 rpm. Not bad for an engine that looks as good as it runs.

The cylinder liners in the aluminum block come unmachined, so the first order of business was to bore (shown) and hone the cylinders to a 4.125-inch diameter.

The aluminum block has thicker main webbing than a cast-iron block at the oil pan rail mating area, so the 4340 Summit crank needed to have the large counterweights on either side of the No. 2 and 4 mains clearanced .200 inch.

The studs for the splayed bolt main caps are provided by GM. They were heavily lubricated and then screwed into place. The bearings are Clevite tri-metal pieces oiled up and installed. Notice the trick screw-in freeze plugs on the block (arrows 1) and the small-diameter oil filter mounting pad (arrow 2). Not shown is the lack of a fuel pump mounting boss.

The main caps are indexed with the block from the factory to ensure they go in straight on the aluminum block. The torquing procedure and specs are the same as on a cast-iron block. The Summit crank has not been balanced yet in this photo—we are just checking to make sure the counterweights are properly clearanced.

sion. There are four-wheel disc brakes doing the whoa under a DuVall windshield, which is bolted to a Harwood fiberglass roadster body on highly modified Deuce Just-A-Hobby framerails. Whew! This thing is sick with tricks!

To top it off, this car is going to be fully functional. The QuadraDeuce will join us on our Power Tour in May, and then it will compete in the One Lap of America competition in June. The QDeuce is the brainchild of Summit R&D chief Mark Stielow, who won the Vintage class last year in the One Lap with his trick '69 Camaro.

This car is designed, and is being built, to be bitchin', but its real purpose is to boogie. Carefully modifying and combining off-the-shelf Summit components with industry standard pieces, the docile 400ci all-aluminum small-block made a peak 520 hp at 6000 rpm with a strong powerband. Even though this engine is a high-buck build, there are a lot of good ideas for making power here, so take a look at how the pros at Summit do it. (And watch future issues for more on the QDeuce.)

Since this engine will see everything from stop-and-go traffic to racetrack loading, the piston rings need to be excellent performers under all situations. Summit chose 1/8-, 1/8-, 3/16-inch Speed-Pro moly faced units, because they are solid performers in high-stress applications.

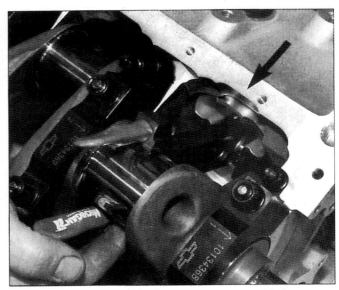

Summit's Pro-Line rods were installed with ARP rod bolts. Notice how the aluminum block comes clearanced for the 3.75-inch stroke crank (*arrow*).

GETTING THE FLOW

The TFS Twisted Wedge cylinder heads make serious power right out of the box for engines close to stock. But the QuadraDeuce 400ci engine is a heavy breather, so the heads need to flow a lot of the air/fuel mixture. While the inlet and exhaust ports were fully modified, the strengths of the Twisted Wedge TFS head design, the small combustion chamber and the overall port shape, combined to help this Mouse churn out over 500 hp.

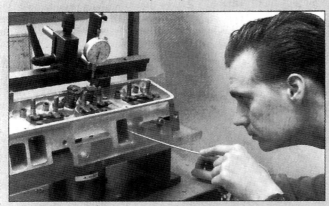

With considerable time on a SuperFlow flow bench, revised intake and exhaust port configurations were determined for the TFS heads.

The optimum exhaust port location required the port to be moved .130 inch up from the TFS location. The bolt holes were welded up and redrilled to mount the headers in the proper location on the head.

The TWM intake was opened up slightly and straightened out to promote better flow and keep the fuel from puddling.

A Brzezinski scribe guide was used to port-match the intake manifold runners to the cylinder head intake runners.

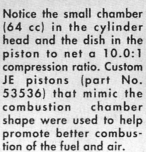
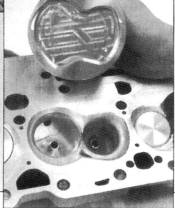

Notice the small chamber (64 cc) in the cylinder head and the dish in the piston to net a 10.0:1 compression ratio. Custom JE pistons (part No. 53536) that mimic the combustion chamber shape were used to help promote better combustion of the fuel and air.

A custom pickup was welded to a high-volume wet-sump oil pump at 3/8 inch off the bottom of a Stef's Fabrication Specialties oil pan for maximum scavenging.

After modifications outlined in the accompanying sidebar, the TFS Twisted Wedge cylinder heads were bolted on with ARP fasteners and a Fel-Pro 1034 head gasket to achieve a tight seal at all surfaces. A Crower roller cam is being used, as are Crower lifters and valvesprings. The cam (part No. 00427) is a street roller with 256 and 262 degrees duration at .050 inch and .585 and .597-inch lift on the intake and exhaust, respectively. Lobe centers are 112 degrees apart. TFS guideplates are required on the Twisted Wedge head as it relocates the pushrods slightly.

A Crane rev kit was modified slightly by opening up some of the pushrod holes in the spring retaining plate to fit the valvetrain configuration on the TFS cylinder heads.

Crane roller-tip rockers, 1.5 on the exhaust, 1.6 on the intake, were used to fine-tune the valve lift for optimum power output.

To ensure the engine can take any rpm thrown at it, one of Edelbrock's new camshaft belt drives is being used along with ATI's Super Damper. The belt drive allows for easy cam adjustment or removal, which will come in handy as the engine combination is dialed in.

With an Edelbrock water pump and other accessories bolted on, the engine is dropped into the QuadraDeuce for some final mockup. The QDeuce ought to be quite a ride with this baby making the steam! **HR**

SOURCES

Brzezinski Racing Products
Dept. HR05
H50 W23001 Betker Dr.
Pewaukee, WI 53072
414/246-8577

Chevrolet
see your local dealer

Summit
Dept. HR05
P.O. Box 909
Akron, OH 44309-0909
216/630-0200

Trick Flow
Dept. HR05
1248 Southeast Ave.
Tallmadge, OH 44278
216/630-1555

TWM Induction
Dept. HR05
325 Rutherford St.
Golita, CA 93117
805/967-9478

454 SMALL-BLOCK
Continued from page 31

Conely guarantees the springs for life against breakage or the loss of more than five pounds of tension.

Since Jack runs his long-stroke engines in the 7500rpm range, he has given the lubrication system much work. The heart of the system is a big-block Chevy oil pump, originally designed for the ZL-1/427. This particular pump has longer gears (1.3 inches overall) than the small-block or standard big-block oil pumps for greater capacity, and Jack makes several modifications for use in his racing engines. The pump's slotted drive shaft is pressed .170-inch deeper into the gear. This insures that the distributor will have adequate clearance to seat properly when placed in the top of the engine. The burrs are removed from the teeth of the pump's gears with a die grinder, and slots are cut in the body casting and cover with an end mill to prevent cavitation of the oil supply and the possibility of pumping only air. The cover is installed with a .0025-.003-inch clearance between it and the gears. Conely stocks the already modified ZL-1 pumps, selling them for about $35. The windage tray must be reworked to clear the body of the larger big-block pump. Jack also cuts slots in the windage tray for increased oil drainback. The slots are located to take advantage of the direction of rotation of the crankshaft, using the wind from the crank to force the oil out of the tray and back into the pan.

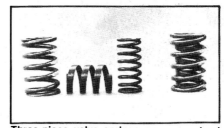
Three-piece valve springs are guaranteed against breakage/loss of 5-lbs. pressure.

Conely details his engines with such niceties as a Mallory magneto and Crower port injectors. He also uses rare, Chevy-built aluminum, small-block water pumps, which he sells for about $25.

On the dyno, the injected 454-inch small-blocks produce over 600 horsepower at 6800 rpm. That's good power in anyone's book. But when you consider that a small-block weighs about 150 pounds less than a comparable big-block Chevy engine, that's like being handed even more free power. Power to make people believe in the small-block.

By Don Green and Terry Cook

Performance-wise, Chevy's new "giant" small-block disproves the old adage, "There's no substitute for cubic inches," but there are still ways to harness a...

400 inch bust

THE PEOPLE AT CHEVROLET HAVE GOT TO BE "NUMERO UNO" AT GETTING THE LONGEST LIFE OUT OF THEIR ENGINE DESIGNS. In the fifteen years they've been producing V8 engines, they've had only three basic designs, two of which are still very much alive. Only the 348-409 engine design has disappeared from the option lists. Of the two remaining engine designs, the undisputed longevity champ is their "small block." Since 1955, it has undergone a multitude of bore/stroke changes and design refinements, having been a 265, 283, 302, 307, 327 and 350-incher.

Now it's 1970, and that same small block Chevy engine is still with us. But the only thing that is small about the small-block is its exterior size. Inside, it's a giant! The engineers at Chevy have bored and stroked that high winding little engine all the way out to 400 cubic inches — a 4.125 inch bore and a 3.75 inch stroke, a considerable increase from the 3.750 x 3.00 inch 265 cube '55 block. Is nothing sacred?

If you've been looking through your Chevy dealer's sales materials, you've probably noticed that there are *two* 400 cubic inch engines in the lineup this year. One, of course, is the new small-block we've been speaking of, with its 2-bbl. carb and 265 horsepower. The other, a 4-bbl./355 horsepower version, is a *completely* different animal. It is basically a low-performance 396 with a slight over-bore, and offers less performance than the standard 396. Here, however, we are only involved with the small-block 400 design.

Known at Chevy plants and dealerships across the country as RPO LF-6, the small-block 400 is a stone! Don't get us wrong, though — Chevrolet intended it that way. They wanted a high-torque, low-rpm engine that would run on regular gas while pushing any of Chevy's heavier entries down the freeway. And they got it!

Performance-wise, Chevy's engineers will be the first to tell you that the LF-6 is not "where it's at," as far as its production components are concerned. For openers, forget the carburetor, intake manifold, cylinder heads and camshaft, as they are all low-performance items that have no racing applications. When we start looking into the short block, however, things begin getting a little more interesting, but only a little. The pistons are conventional cast, low-compression production items, and are extremely heavy when compared to the current crop of racing pistons. But they are ideal for the 400's designed purpose. At low rpm, they should last forever. Give the 400 a good high-rpm cam and some carburetion and the increased rpm would pull those same stock pistons apart.

The connecting rods are forged, but are made of the same 1046 steel mate-
(continued on following page)

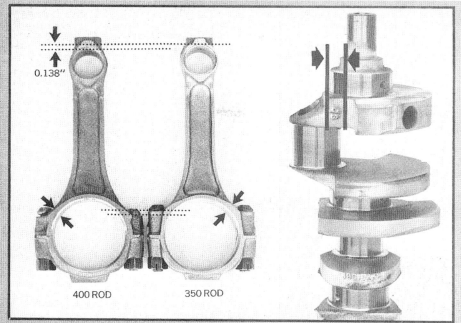

FIG. 1 — 400 rod is shorter in overall length; bolt boss is shorter and weaker.

FIG. 2 — Large 2.650-inch main bearings overlap with rod journals adding strength.

FIG. 3 — Small holes between cylinders carry water from block to heads, and must be plugged when using high-performance (non-400) small-block heads — tapered brass or lead plugs can be used; large bore causes "siamesing" of adjoining cylinders.

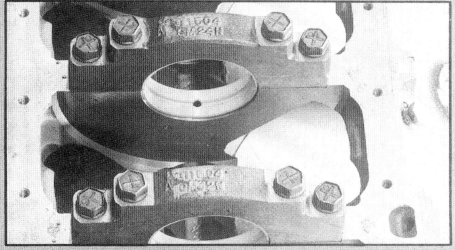

FIG. 4 — All 400-inch small-block Chevys are produced with the proven 4-bolt mains.

FIG. 5 — Center crank is stock 400 unit; at right is the original rough forging; at left is the counterweighted CSC destroker.

FIG. 6 — CSC modifies stock, rough rod forgings for added durability; includes polishing of sides to reduce cracking.

FIG. 8 — Stock cast piston is shown with CSC's forged racing piston having unique "head-land" rings and "Fire-Slot" dome design; stock pistons are extremely heavy.

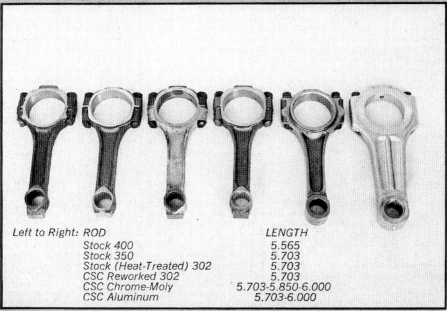

Left to Right: ROD	LENGTH
Stock 400	5.565
Stock 350	5.703
Stock (Heat-Treated) 302	5.703
CSC Reworked 302	5.703
CSC Chrome-Moly	5.703-5.850-6.000
CSC Aluminum	5.703-6.000

rial used in all of Chevy's standard, low-performance production rods. And even though the 400's stroke has been increased a little over a quarter-of-an-inch from the longest production small block crank previously available from Chevy, the connecting rod length is actually shorter than any previous small block rod. The center to center length for the 400's rods is 5.565 inches, while all other small block Chevy rods are 5.703 inches long. (figure 1) If you recall our *Tech Series* article on connecting rods in last month's issue, it was pointed out that long rods, rather than shorter ones, is the plan. The shorter rod creates a severe rod angle, increasing the piston's side thrust and causing additional (undesirable) cylinder wall loadings. But it is also a characteristic of the shorter rod to produce increased low rpm torque. With that in mind, it is easier to understand Chevy's thinking behind the shortening of the small-block rods. While there may be some oddball custom engine situation which requires a shorter rod (we can't think of any, offhand), the short 400 rod is otherwise useless for performance applications. Another bad point is that the bolt bosses on the 400 rod have been shortened. When compared to the other late-model small block rods, this represents a loss of strength in the critical area near the bolt head (figure 1).

The 400-incher's crankshaft is also a departure from other late-model small-block Chevy cranks. While everything from the 250 hp/327 on up has been produced with a forged steel crank, Chevy has switched back to nodular cast iron for the 400's crank. Again though, the crank is perfectly suited to its designed function — low-rpm operation. At 3.75 inches, it is the longest stroke crankshaft ever produced by the factory for the small-block engine family, and provides a tempting piece of raw material for a moderate performance street machine. For use in an earlier (non-400) block, the new, larger-than-ever 2.650 inch diameter mains must be machined down to the usual small-block sizes of 2.450 or 2.300 inches. These larger mains have two effects on the crankshaft and its usage. Larger mains mean larger bearings and more bearing surface, and to a racer that means more bearing speed, more potential friction and more drag. For that reason, it is extremely common to find racers using the pre-'68 block and crankshaft assemblies with the smaller 2.300 inch diameter mains, which understandably have less bearing drag. But the large mains offer increased rigidity to the crank, as the diameter of the mains overlaps with the diameter of the crankpins at the ends of the throws (see figure 2). This provides a stiffer crank and reduces the torsional vibration during operation, resulting in less flexing and cracking. For an engine of this size (cubic inches) and considering the horsepower potential, the benefit of the added crank strength may offset any disadvantage of the increased bearing surfaces.

Like all the "big-block" 454's, the 400 inch small block is externally balanced. Rather than making the balance compensations on the crankshaft itself, Chevy has used the damper on the front of the crank and the flywheel on the rear to control the balance. While the husky 454 cranks are rigid enough to handle it, the more spindly 400 crank becomes subject to a great deal of flexing by having all the balance weights at the extreme ends, even with the overlapping main and rod journals.

While the 400 is made up of many new parts, none of the ones we've mentioned so far really lend themselves to high-performance work. But we've saved the best part for last — the cylinder block. The LF-6's 4.125 inch bore has never been possible before with any of Chevy's small blocks (without extensive sleeving). To make it possible for the 400, Chevy re-cored the casting, providing so much material around the cylinder bores that the metal actually flows together between each cylinder producing what are referred to as "siamesed" cylinders (see figure 3). There is no longer any provision for water to flow between the cylinders — that area has become a web of solid iron, but the web produced by the siamesed cylinders also adds a good deal of strength to the block, helping the cylinders resist flexing. Chevy has left enough material in the casting around the cylinder bores to easily allow a 0.030 inch overbore, while some have been successfully bored as much as 0.060 inches over stock, though 0.060 over is not recommended due to the lack of suitably large head gaskets. And all this with Chevy's already race-proven 4-bolt mains (figure 4).

So we now have a cylinder block that offers us something never before available from a factory-produced small-block Chevy — a huge 4.125 inch bore. And since the factory has designated it as a non-performance engine for the present, it remains for the speed equipment industry to carry the ball. And one of the men who has quickly begun a performance development program based around the LF-6 is Hank Betchloff of the Crankshaft Company in Los Angeles. Hank has already gone into production on a line of forged crankshafts to *destroke* the 400-incher, taking advantage of its big bore and giving it the kind of stroke that will let the small-block do what it has always done best — go for that high-rpm (by reducing piston travel). The CSC cranks are available in three basic strokes: 3.000; 3.250; and 3.480 inches, comparable with the 283, 327 and 350 stock cranks, and all have the large 2.650 inch mains. All three use the 2.100 or 2-1/16 inch rod journals. Working with an LF-6 block with a 0.004 inch clean-up bore, the possible displacements are 321.3, 348.1 and 372.7 cubic inches. What's so unusual about that, you ask? True, you've been able to get very nearly these same displacements from small-block Chevy's long before now, without the 400 cylinder block, but now you're getting these displacements with a full one-quarter inch *shorter* stroke than ever before.

Now do you see the light? That 350 cubic inch small-block you've been racing with its 4.00 inch bore and 3.48 inch stroke can now be replaced by a CSC equipped small-block with 348 cubic inches from a 4.129 inch bore and a much shorter 3.250 inch stroke. Or bore the block 0.030 inches over and get 352.5 cubes with that same short stroke. If you've been running a 327 with a 4.00 inch bore and 3.25 inch stroke, CSC can fix you up with a 321 inch engine with a quarter-inch shorter 3.000 inch stroke. Or bore the 400 block 0.030 over and the 3.000 inch crank will give you 325 cubic inches. And who can doubt that the CSC setup is going to be good for a gob more rpm than the stock 350 or 327.

For the guy who is looking for a particular cubic inch size engine to give his car the best break in a certain class, CSC reworks the basic short stroke cranks to offer stroke variations amounting to plus or minus about five cubic inches from the basic cranks. This gives you a ten or eleven cubic inch latitude in your engine's displacement, and is made possible by grinding the crank pins undersize to 2.000 inches and slightly off center. If you want a few *more* cubic inches, you grind the new rod journal to the *outside* of the original journal; if you want a few *less* cubes, grind the journal to the *inside* of the original. So, with a 4.129 inch bore, CSC can vary the stroke of the basic 3.250 inch crank to yield from 316 to 326.7 cubic inches. Likewise, the basic 3.480 inch crankshaft, when used with a 0.060 inch overbore of 4.187 inches, can be modified to give a total displacement varying between 377.8 and 388.2 cubic inches.

If all you want is just a good moderate performance crank based around the stock 400's 3.75 inch cast crank, CSC can give it their standard "Ultra-Rev" treatment. This includes cross-drilled mains, fully radiused mains and throws, and with optional Tuff-Triding, will provide a suitably strong crank for a moderate performance, high-torque street engine. But for all-out high-performance use, CSC has the *short stroke* forged cranks available either in their fully-counterweighted "Dyna-Rev" style or their non-counterweighted "Ultra-Rev" (figure 5).

FIG. 7 — Stock cast 400 piston (left), and CSC's forged flat top racing piston.

FIG. 9 — Head-land rings are of dykes design; placed at extreme top of piston sides; good low and high rpm sealing.

CSC can also provide the longer rods needed to make the engine perform well in the upper rpm ranges without changing displacement; anything from mildly reworked stock rods to their own super-duty, forged, chrome-moly steel or forged aluminum rods. In the 2.100 inch crank pin size, they rework the 5.703 inch long 350 cube Chevy rods, selecting the hardest ones after Rockwell testing and magnafluxing; all rod bolt surfaces are milled parallel to the rod cap parting line; rod bolt holes are rebored parallel and enlarged to use 3/8 inch o.d. bolts rather than the stock 11/32 inch bolts; an even radius is cut around the rod bolt head and nut to relieve stress; and the forging lines are removed from the rod and the sides polished in the proper lengthwise manner (figure 6) followed by a complete

Continued on page 43

BRACKET BUILDUP

Building A Big-Cubic-Inch Bracket Race Engine Doesn't Have To Cost Big Bucks

By John Kiewicz

As car crafters, we all love horsepower—the more power, the better. But to build big horsepower, usually you've got to shell out big bucks for good parts. A custom crank here, a one-off cam there; it all adds up to tons of torque, but at the expense of a lightweight wallet.

What if we told you that we've discovered a competitive bracket race engine that not only produces more than 450 horsepower, but costs about $4200. Right about now you're saying, "Oh sure, but the kit probably only comes with a crank and rods—the heads, intake, cam and other components are optional."

We're here to tell you that one company, California Discount Warehouse (CDW), has an engine buildup package that includes a machined block, crank and rods, as well as brand-new forged pistons, cam, lifters, cylinder heads, roller rockers, intake manifold and so on. We're not talking cheapo parts either; how about goods from Edelbrock, Comp Cams, Ross, Dart and Fluidampr—just to name a few.

With the Cal Discount package, you get all of the engine goodies except for the buildup know-how. But that's where CAR CRAFT comes in; we offer insight and photo guidance on how to build the bracket race bruiser. Follow along closely and you'll be doing smoky burnouts at the dragstrip before you know it. And trust us, the CDW big-cube 415 has enough grunt to launch your car down the 1320 and possibly even into the winner's circle.

Directly after building the engine, California Discount Warehouse bolted the V8 into a 3350lb '67 Camaro bracket race car and ran consistent e.t.'s of 11.20 at about 122 mph! In fact, the combo was consistent enough to garnish the First Place trophy at that evening's run-what-cha-brung drag race event. No kidding. In any case, building the CDW 415 small-block Chevy is fairly easy, but a few key items need to be addressed—so pay attention and hold on as we launch into our bracket buildup.

The heart of the Cal Discount Warehouse 415-cube bracket race engine is the bottom end kit (PN SS415KF). Included within are Ross forged aluminum pistons, TRW rings, reconditioned rods, a custom-ground crankshaft and Michigan 77 bearings.

One of the keys to the CDW 415 combo is the specially offset-ground crankshaft. The crank started life as a 3.75-inch-stroke 400 small-block unit with 2.10-inch-diameter journals, but the rod journals have been offset-ground and machined to a 2-inch diameter. After machining, the stroke is now 3.832 inches. Before installing the crank, always measure *every* journal to ensure consistent and correct specs.

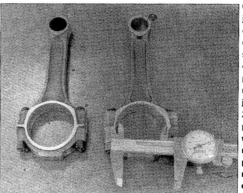

CDW uses small-journal 327 rods in conjunction with the offset-ground crankshaft to arrive at the 415-cubic-inch displacement. The 327 rods have a smaller journal diameter of 2 inches (as compared to a 400 rod with 2.10-inch journals) that mates perfectly with the reground 400 crank. Just be sure to use the correct bearings (included) that will give the proper clearance. In the photo notice the difference between the journal size of a 2-inch-diameter (5.7-inch length) 327 rod *(left)* versus a 2.10-inch-diameter (5.565-inch length) 400 rod.

BRACKET BUILDUP

Because of the different rod/stroke ratio of the CDW 415 V8, to gain proper clearance you must partially grind the rod bolt to clear the camshaft as the rod/crank combo rotates through its firing sequence. Cal Discount machines the bolts for you (part of the kit price), but you must remember to install the rods correctly on the crank to take advantage of the "clearancing." And yes, the rods were balanced *after* the bolts were machined.

Installing the CDW crank is identical to the installation of standard cranks—just be sure not to nick the journals on the main studs (optional) when lowering the crank into place. Notice the small rubber caps placed over the main studs. We recommend these bolt covers from Michigan 77, because they help protect the crank from damage (scrapes, gouges and so on).

Before installing the main caps, be sure to apply a liberal coat of lube to the bearings. Never install the bearings dry, or gouging and/or seizing will result.

Because bracket race engines see a lot of abuse, be sure to check all critical clearances such as crankshaft endplay. The endplay varies depending on whether the vehicle will run a manual or automatic trans, what size/rpm converter will be used, as well as other factors, such as if nitrous oxide will be used. For this engine, CDW set the endplay at 0.008 inch.

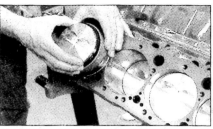

The Cal Discount 415 V8 uses Ross 4.155-inch-bore (0.030-inch overbore 400ci) pistons. The flat-top pistons will net a final compression ratio of 11:1 when paired with 64cc combustion chamber cylinder heads. Notice the ring compressor. CDW sells this handy tool that allows you to reduce or enlarge the diameter, making it perfect for installing pistons of various diameters.

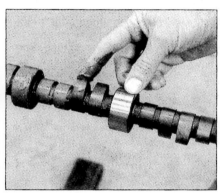

The CDW 415 uses a fairly lumpy solid lifter camshaft. The Isky 275 cam features 0.570-inch lift with 274 degrees duration at 0.050-inch lift (320 degrees advertised duration). The cam is ground with 108-degree lobe centers, which gives added clearance to the rods as they pass by the cam lobes during operation. To maintain proper clearance, a 110-degree lobe separation camshaft is the widest that can be run—otherwise rod-to-cam contact will occur. A small base circle cam is not needed. See illustration below for more info on lobe separation. As with any non-roller camshaft, before installation, always apply plenty of cam lube to prevent the lobes from going "flat."

The reconditioned rods come with stock rod bolts, but for an extra 50 bucks CDW can upgrade to ARP chrome-moly rod bolts. Be sure to torque the $11/32$-inch rod bolts incrementally up to the final 40 lb-ft setting. Once all of the rod and main bolts are torqued to spec, slowly rotate the crank assembly to ensure that all is correct and nothing binds.

Most novice engine builders install the rods and never check the side clearance. Because bracket race engines operate under extreme conditions, obtaining the proper side clearance is critical. Always check clearance with the rod and main bolts torqued to spec. On this engine, the side clearance was 0.012 inch, which is fine. In general, never run clearances tighter than 0.070 or looser than 0.180 inch.

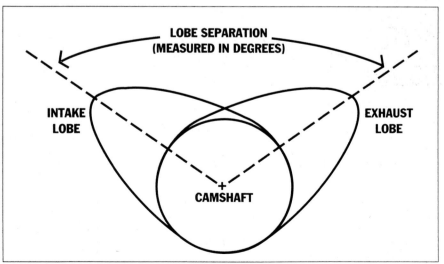

Lobe Separation

Camshaft lobe separation is the distance, measured in camshaft degrees (camshaft degrees equal crankshaft degrees divided by two), between the centerlines of the intake lobe and the exhaust lobe. Tighter lobe separation increases intake and exhaust lobe overlap, which usually generates better low-end power but sacrifices smooth idle characteristics and manifold vacuum.

BRACKET BUILDUP

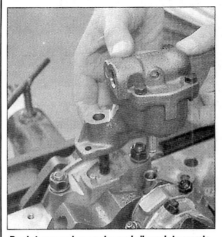

Bracket race engines need a good oil supply to operate properly at high rpm. Thus, CDW includes a Melling M55 high-volume oil pump with the 415 engine package. The pump also comes with a pickup, but be sure to measure pickup-to-oil-pan spacing to ensure adequate clearance.

To ensure that there will be plenty of oil during the wheels-up dragstrip launches, CDW includes a custom seven-quart oil pan in the engine package. Even though the pan is a low-profile race-type unit, its low price will please you. The cost is $85.99 if bought separately.

Included in the CDW 415 stroker engine package is a pair of Dart Sportsman cylinder heads. The heads feature all of the hi-po goodies such as hefty valvesprings, hardened exhaust seats, screw-in studs and revised runners for better flow. Then again, if you desire even better flow, for an extra charge Cal Discount can do all sorts of port and polish work. Call for specific prices. The heads come standard with 2.02/1.60-inch valves, but if you want 2.05-inch valves add $100 to the tally.

When installing the Dart Sportsman cylinder heads, be sure not to damage the head gaskets in the process. If so, blow-by and/or coolant leaks may occur. To ensure a proper seal, CDW includes top-quality Fel-Pro head gaskets (part of a complete Fel-Pro kit) in the engine package.

Most engine kits don't include items such as rocker arms, and if they do, they're usually stock stamped-steel-type units. The CDW bracket race engine kit comes with Competition Cams 1.52:1 roller rocker arms that can handle the high-rpm abuse the engine will endure. With a rollerized fulcrum and tip, the Comp Cams rockers reduce friction, freeing up extra horsepower.

Capping off the 415 brute is an Edelbrock Victor Jr. aluminum intake manifold. This high-rise, open-plenum manifold will provide plenty of breathing as the big cam beckons for lots of air/fuel mixture. Although the redline on this engine is in the 6500rpm range, the manifold is good for 8000 rpm! Notice the cool-looking polished aluminum valve covers from Cal Discount Warehouse. They aren't included in the engine package, but they're so low-buck ($59) that you may wish to add a pair.

Here's the big-brute 415 in its completed state just before it was taken to the dyno facility for flogging. Notice the harmonic dampener. Included in the 415 kit is an 8-inch Fluidampr that will thwart harmful high-rpm harmonics. Although this engine has a polished aluminum timing cover and valve covers, they don't come as part of the package. But good news; they're super inexpensive—the timing cover is only $45.99, for example.

Once the engine was built, California Discount Warehouse took it to Westech Performance to have it dyno-tested. After several engine pulls on a Superflow dyno, the best horsepower output was reached when running 34 degrees of total ignition advance along with No. 77 jets in a Holley 750cfm double-pumper four-barrel carburetor. All dyno testing was done through Hooker Competition Plus headers. In fact, one dyno pull was made running through a pair of Flowmaster 3-inch mufflers—the engine output was only down by *five horsepower!*

California Discount Warehouse 415 V8 Kit

**Prices as of March 10, 1995

Bottom End Kit (PN SS415KF)
Custom-offset-ground crankshaft
Reconditioned 5.7-inch rods
Ross forged pistons
Sealed Power piston rings
Michigan 77 rod and main bearings

$849.99

Buildup Parts For CDW 415 Engine

Fully machined cylinder block	359.99
Edelbrock Victor Jr. intake	199.99
Dart Sportsman cylinder heads	899.99
Comp Cams Pro Magnum rocker arms	229.99
Isky 275 cam and lifter kit	149.99
Manley cam lock	3.20
Manley cam button	8.75
CDW chrome-moly pushrods (hardened)	49.99
CDW Tru Roller timing chain set	29.99
CDW aluminum timing cover	45.99
CDW aluminum valve covers	59.99
CDW head stud kit	79.99
Pioneer main stud kit	34.90
Fel-Pro Perma Torque gasket set	39.99
Fluidampr harmonic balancer	329.99
Pioneer SFI-approved flexplate	64.99
CDW 7-quart oil pan w/pickup	85.99
Melling M55 high-volume oil pump	24.99
Oil pump shaft	9.70
System 1 oil filter	65.99

$2774.39

Additional Mods (add to price of kit)

CDW Stage 2 cylinder head porting	300.00
CDW 2.05-inch intake valves for heads	100.00
CDW electronic balancing of parts	85.00
ARP chrome-moly rod bolts	50.00
Sealed Power moly piston rings	15.00

$550.00

Total Cost Of CDW 415 V8

Bottom End Kit	849.99
Buildup Parts In 415 Engine	2774.39
Additional Mods	550.00

$4174.38

Sources

California Discount Warehouse
Dept. CC
2320 E. Artesia Blvd.
Long Beach, CA 90805
310/423-4346

Westech Performance
Dept. CC
11098 Venture Dr., Unit C
Mira Loma, CA 91752
909/685-4767

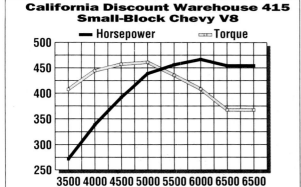

Continued from page 39

shotpeening (rod and cap). Also available are the factory heat-treated 302 rods in both the 2.000 inch and 2.100 inch sizes (5.703 inches long). CSC's forged chrome-moly rods can be had in three lengths: 5.703; 5.850; and 6.000 inches — all in either 2.000 or 2-1/16 inch pin sizes. The forged aluminum rods come in either 5.703 or 6.000 inch lengths.

The CSC forged pistons are available in a full range of bore sizes from stock to 0.030 inches over, (0.060 over on special order) and pin heights ranging from 1.800 to 1.250 inches, as measured from the center of the pin to deck of the piston. Both flat tops and the CSC "Fire-Slot" design domes are available (figures 7 and 8), and both can be made to take advantage of a unique ring design made possible for the small-block engine by the 400's larger bore size. Called "head-land" rings, they are placed in the top ring groove which is cut just below the piston's deck surface (figures 8 and 9). The ring is of a dykes design; high ring placement fully exposes the ring to the pressure in the cylinder, making it work more effectively. It has not been possible to use the step-land design with some of the smaller bore Chevy pistons because the valve clearance notches were placed so close to the piston edge that the high, top-ring groove cut through into the clearance notch. The 400 incher's pistons have enough material around the outside of the piston to cut a satisfactory ring groove without worrying about cutting into the valve notch. A coil expander keeps tension on the dykes ring, forcing it against the cylinder wall to keep a good seal at low rpm.

Chevrolet obviously does not *need* a high performance small-block engine in the 400 cubic inch range, because everything that is wrong with the small-block 400 from a performance angle, is right with their 396 or 400 inch big-block "Rat" motor. Even weight is not a factor when you consider that the ZL-1's aluminum block is now readily available — if you have the *price*. But the basic design of the small-block has been around long enough that there is a ton of equipment, new and used, just waiting to be bolted to a 400 block, be it stock, stroked or destroked. For a non-performance engine, this one is bound to run!

"Well, it saved me the price of a paint job..."

LOOKING INSIDE a 200 mph CHEVY

By LeRoi Tex Smith

SOME NAMES FLASH on the hot rod scene like a thunderclap, arriving out of nowhere. And, often returning just as fast whence they came. Others just seem to show up one day and take an expected place at the head of the order. Jack Lufkin is such a fellow.

Jack is no newcomer to hot rodding. He started insignificantly enough, with flathead engines and sportsmen class circle cars, back in Massachusetts. In those days, you didn't have much of an advantage over your pit neighbor, so you either got very good in the business or got nowhere. Jack got good.

After a stint in the service, Lufkin showed up in Southern California. Before long, he was working for the old man himself, Ak Miller. An association that just naturally led to some competition at the dry lakes and Bonneville. Actually, Jack first ran a car at Daytona for the flying speed record attempts, and did right well for himself. His first wail at the lakes came in 1959, with a modified '56 Corvette. This initial attempt was rather fruitful for Jack; he stills holds the B/Sports record at Russeta and SCTA lakes meets and at Daytona.

Then, in 1963 this tall, quiet fellow from the East Coast made the boys really sit up and take notice. He cranked a stock-looking '62 Corvette through the timing lights at Bonneville National Speed Trials at 193-plus mph. Unblown!

But then, the performance didn't surprise the people who know Jack Lufkin. Because, this is one man who prepares for the adventure. Nothing is left to chance. Nothing is slipshod. Especially in the engine department. But, what kind of an engine is it that can be run normally aspirated at nearly 1 mile altitude, on gas, at such a fantastic speed? Let's take a look at how Jack builds a Chevy engine.

While spending some time at Miller's shop recently, checking on the progress of Ak's latest Pikes Peak hauler, I noted a new engine going together in the back of the building. Questions revealed it to be another of Lufkin's creations, being built to compete in C class at the Salt. The chassis is to remain the old reliable (1 year old, that is) '62 'Vette. But instead of being satisfied with just 193-plus mph, it's all-out this time, with 200 the goal.

This new engine incorporates all of the Lufkin basics and secrets to success. The basic block is a '64 Corvette. Amazingly enough, very little has been done to it.

The first step in the modification procedure was to completely disassemble the engine, carefully inspecting each piece and keeping everything in absolute order. And

You don't have to have a huge supercharger, a can of nitro and mucho cubic inches to go fast. Just look what Jack Lufkin accomplishes with a medium sized Chevy

These are the basic essentials necessary to convert a common Corvette engine into a fire-breathin', salt stompin' charger. Of course, it helps the entire project if the builder knows the score.

clean! Jack is a bug on cleanliness (quite apparent around the shop) and insists that cleanliness is next to godliness (except in Texas, where it's next to impossible, as any ex-serviceman knows).

The block should first be align-bored. This will insure that all of the main bearing webbs are in alignment. This is the first place where many an engine builder falls off. It is absolutely worth the extra effort, as a slight misalignment will tend to bind up the crank just enough to retard completely free rotation. And, this condition can sometimes lead to an unscheduled disassembly of the engine. Usually at speed.

The cam bore is also checked for proper alignment. Lufkin had his work done by C-T Automotive. Clem Tebow is an old hand at super performance engines, and this basic preliminary check is standard procedure.

Next, the block cylinder bore is checked. Because all engines are mass produced, there is bound to be a slight deviation from perfect. The blocks are bored by a gang of tools. On a high performance Chevy, always check and see that the bore is absolutely perpendicular with the head mating surface *and* the centerline of the crank. This relationship should be a perfect 90 degrees, nothing else. Usually, any deviation will be found in the head mating surface. On this particular engine, one of the surfaces was .022 while the other showed .026, .004-inch deviation from one end to the other.

Now is the time to do any valve relief work to the top of the cylinder bore. If you're using a very tall cam, and stock Corvette big valves, there will be interference right at the edge of the bore. Place the heads on the block. From the bottom side, scribe a line on the heads duplicating the exact bore margins. Where the edge of each valve comes nearest to the cylinder wall, a maximum ⅛-inch relief may be cut. This is hardly more than a slight rounding of the bore lip at this point, but it means the difference in bent valves, or not.

C-T Automotive also did the crankshaft for this particular job. Since the C/Sports Racing class limits are 375 cubic inches, unblown, the bore was left stock. Instead, C-T took one of their special 3¾-inch stroker cranks and destroked it several thousandths to produce a total of 374.97 inches. How's that for getting right on the gate post? Of course, many a would-be record holder at the salt has been disqualified because of a slight overabundance of inches, so Jack has checked this engine size thoroughly.

The crank has had .0025-inch clearance established on all bearing surfaces. Rod end play has been kept at the high limits of stock specifications. Standard end play of the crank at the rear main is set at .007-inch.

(continued on following page)

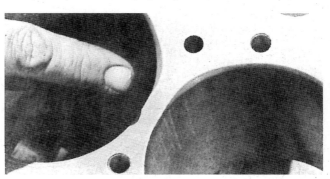
One of the early block preparation steps is relieving the top of the cylinder bores for large valve head clearances. On a late Chevy block, ⅛-inch is about the maximum grind. Don't go over.

A stroked crankshaft is used in the Lufkin engine, this one being set up by C-T Automotive with larger oil grooves on the mains and necessary balancing plates added to the outer counterweights.

After the block is align bored and thoroughly cleaned, the crank is carefully lowered into position and checked for rotating clearance. It's been found that engines need slight block grinding.

200 MPH CHEVY

A substantial assembly lubricating agent is applied to the main bearings, then the caps locked into place. After torqueing the cap screws down tight, check to be sure that crank spins freely.

With the crank securely in place, the clearance of the thrust surface at rear main is checked. This clearance is often overlooked, and can mean the difference of winning or losing in competition.

With the pistons temporarily in place on rods, check each one in proper cylinder. Jack always uses the recommended piston manufacturer's clearances and doesn't take even smallest chance.

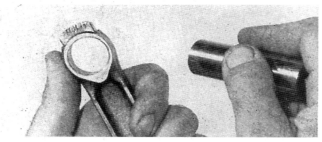

Normal 327 rods were magnafluxed, set up to accept the special Forgedtrue wrist pins. The pins are fitted to a close tolerance, while leaving them free enough to move in the bushings easily.

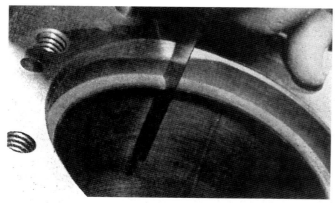

The end gap of each and every ring is checked in its respective bore. This is a ritual with Lufkin, and is one of the reasons for his outstanding engine performances and life. This is important.

After the crank was returned from C-T, it was thoroughly inspected and all minute burrs and flaws removed. Most speed manufacturers caution the user to take the time for this effort, but all too often the novice builder overlooks this vitally important point.

Lufkin doesn't set his engines up nearly as loose as some builders, but this might be the exact reason he gets such life from them. He uses 30 weight MS high detergent oil, recommending any good name brand.

Lubriplate is spread lavishly on all the main bearings and the crank is carefully dropped in place. The caps are installed and torqued down to 70 pounds. The end play clearance is checked with a feeler gage, and the crank is spun in its cradle to make sure it spins freely. If the crank doesn't spin free, something is wrong. Don't go further until this point is checked out.

The block-crankshaft assembly is ignored for awhile as the piston and rod are put together. Jack uses Forgedtrue solid skirt pistons. These pistons are truly forged out of a billit of solid aluminum alloy. Forgedtrue and Merriman are the only true forged pistons I know of. The other pistons called forgings are really pressure castings. Anyway, the Forgedtrue slugs have the high dome giving 11 to 1 compression ratio.

Stock 327 Chevy rods are used, after the magnaflux treatment. The upper end of the rods are modified to accept brass pin bushings, also supplied by Forgedtrue. Jack notes that for years he just used a floating pin fit without the bushing. But the pins were always apt to gall. No trouble on that count, now. The wrist pin itself is the special, high-performance item from Forgedtrue. The pin fit is set on a Sunnen hone, quite loose. This is a job best left to the local machine shop.

In the past Jack has had some bad experiences with broken pin locks. So, now he uses the new Ansen Teflon button keepers. These keepers fit right in the recess at each end of the wrist pin and ride against the cylinder wall ever so lightly. The Teflon doesn't wear readily, and will not mar the wall.

The piston-to-cylinder wall clearance is a vital point, and should be checked closely. With the piston dropped in the bore — minus rings — use a feeler gage to make sure there is sufficient room. The piston manufacturer recommendation should be followed. Lufkin uses .009-inch clearance, measured 90 degrees from the pin, as these are cam ground pistons.

Two different makes of rings are used in this engine. Grant rings are used in the two bottom lands, while the special Forgedtrue stainless "Dike" type compression ring is used at the top. This particular ring is now in wide use by special builders. In operation, the piston land is vented to actually let the compression of the cylinder enter behind the compression ring and force it against the cylinder wall.

I remember the first flathead engine I ever built. I just dropped the rings on the piston and pounded the whole shebang in the hold. Lasted until I fired it up. I hadn't checked the ring end gap. This is a must! First, compress the ring and slide it in the bore, squaring it off by flushing it against the top of the piston. Take the piston out again and measure the gap between the ring ends with a feeler gage. On this particular engine, the top ring gap is .032-inch. Number 2 and 3 rings each have .016-inch clearance. Generally speaking, the top, or compression, ring should have .008-inch clearance for every inch of bore. The other rings should have .004-inch. Check every ring in every cylinder bore.

The rings should be placed on the engine with the gaps

After the rings are checked in the bore, they are installed on Forgedtrue pistons. Grant rings are used in the two bottom grooves, with the special new Forgedtrue stainless "Dike" at top.

Rather than take a chance on broken pin locks, Jack uses the new Ansen Teflon pin buttons. They do not wear, and will not score the cylinder walls. Note use of Lubriplate on the bearings.

Photos by Al Paloczy and Tex Smith

staggered. That is, don't line the gaps up, or you got blowby, man. Also, stagger the gaps so they do not fall over the wrist pin. This little trick alone is worth a few extra horses when the chips are down.

With the bearings placed in the rods, the pistons are liberally coated with oil. A Snap-On ring squeezer is pulled down tightly around a piston, and the assembly is slid down into the bore. Make sure that the piston assembly is put together correctly, and that the assembly is then properly placed in the bore. The Chevy units are not marked, but the heavy side of the big rod end goes to the rear on the right bank (cylinder numbers 2, 4, 6, 8) and to the front on the left bank (1, 3, 5, 7). Make sure the rods are in the holes they came out of, and that the eyebrows on the pistons are toward the top. There is usually an arrow on the pistons to direct the builder, and the spark plug relief is another indication.

When the piston is properly selected for the right bore, the ring squeezer is tapped around the top edge to make sure that the piston is absolutely square with the bore. On a reliefed block, the next step of tapping the piston into the bore is critical. If done improperly, cracked rings result. The wooden part of the hammer (butt) is used to gently tap the piston into the bore. While this is being done, make sure the bottom end is guided carefully onto the crank throw.

The rod cap is installed on the correct rod it fits. Some engines are not marked, so be careful. In fact most builders take the time out during disassembly to mark the rod cap and its corresponding rod.

The bearings should be checked before everything is bolted up. With the bearings installed, the rod cap is tightened on the rod. An inside micrometer reading is compared with the size of the rod journal on the crank. After the rods and pistons are assembled in the engine, check the rod size clearance. With the two rods on each journal (V8) slid to one side, there should be room for a .014-inch feeler gage.

After all of the pistons and rods are installed, the deck height is checked. This is the height of the piston top in relation to the head mating surface of the block. On aluminum alloy pistons this should be right on .020-inch. A regular tool is made for checking this height. The pistons are rotated until absolute T.D.C., then a depth mike is used.

Back at the bottom end, make sure that everything clears ok. Strokers usually clear the new Chevy engines, but some grinding is necessary on the older blocks. Naturally, this grinding should be done before anything else.

The camshaft is next. Jack uses a special experimental grind Iskenderian grind along with the complete coordinated

A standard type ring squeezer is used to compress the rings, after they are staggered around the bore. Compressor is carefully squared with the block face, piston tapped gently in place.

After the rods are bolted to the crank, the deck height of the pistons is carefully checked. It is usually too much on standard blocks; however, the head surface may be ground to true it up.

There isn't much room left for piston on a stroker. The Forgedtrue pistons are of the slipper type, with a small tab on each skirt to help resist rocking. Tab must always clear crank well.

kit. First, Rev-Lube is smeared on the lobes. Then, the cam is *carefully* slipped through the bore. Do not scrape or nick the cam bearings. Installed, the cam must spin freely. Using the matching marks on the timing gears, install them.

The tappets are next dropped into place. Jack uses flat tappets, but whatever is used, this is the next step.

Now we go to the heads. Lufkin is using '64 Fuel Injection Corvette heads for the obvious reasons. The ports are big, 2 inches all the way, and stock, lightweight Corvette valves are 2 inches on the intake and 1 9/16 plus .030 on the exhausts. These are the maximum usable sizes without offset guides.

First, the heads were checked for cracks and flaws, then sent out to Jocko's Porting Service. There they were ported and flow tested and the combustion chamber cleaned up. The valve seats are at 45 degrees. The valves have been lapped in, with just the very outer edge seating.

200 MPH CHEVY

If you are building a similar engine with a similar cam arrangement, you'll have to machine the valve spring cups in the heads out to Olds spring size. In addition, a high performance engine should have the rocker stud stands machined off ¼-inch. Screw-in studs are a must, as stock press-fit studs will not stand the increased spring pressures. The amount machined off is for the screw-stud lock nut. Jack uses Isky studs and forged rockers.

Dual springs are used, along with vibration dampeners and alloy retainers. The spring height is set to the recommended (by cam grinder) height. This is done by installing the valve in the head and then slipping on the retainer. With the valve closed, measure the distance from the bottom of the retainer to the top of the spring cut in the head. Under no circumstances is coil bind allowable! Usually, you will find that washers will be needed under the retainer to get the correct height.

It's now time for another clearance check. With number 1 piston on top dead center, install the head with the gasket in place. After checking the rocker arm clearance, adjust if necessary. Rotate the engine 360 degrees exactly. With the cam timing overlap split (measured by a dial indicator), the piston should be at T.D.C. Now, carefully adjust the adjusting nut until the valve just touches the piston. Remember that 1 complete turn of an adjusting nut is .047-inch. Most cam grinders recommend about .125-inch clearance between piston and valve with roller tappets and .050 with flat tappets.

Every valve should be checked this way, since 15 minutes spent here can mean the difference between go and whoa, fun and loss. In addition, the rocker arm contact with the valve tip is checked. The contact point should be right in the middle of the valve stem tip. Contact on either side means excessive side loading and accelerated part failure. One quick method to check this is by sliding a piece of paper between the rocker and the valve stem. Rotate the engine. If the paper comes out torn, the rocker is wiping too much and is not centered. Again, check for coil bind.

Jack uses K-W Copper Coat seal on the head gasket, for insurance. The head is torqued down to 60-70 pounds. If you're going to have the heads on and off often, steel washers under the cap screws will help keep the casting from chipping.

The oil pump should be taken apart, cleaned, and inspected. Look for chipped pump gear teeth. Lubricate the pump and reassemble. After the cam is degreed in (a practice Lufkin follows

Part of the secret of stock 'Vette go lies in big head ports. Jack had Jocko open them up, and standardize every opening.

religiously) and overlap split (he does not lead or retard his cams), the front timing cover is installed. Then the Corvette pan is bolted up. This pan has a main bearing baffle to keep oil splash down to a minimum.

Before anything else is done, Jack primes the engine with oil. This is done by inserting a long screwdriver shaft in a ½-inch drill and running it down the distributor opening to the oil pump drive. With oil in the pan, the drill is run until oil squirts out the oil pressure hole in the block. This way, the engine is not started on dry surfaces.

Lufkin uses Hilborn port injectors that have been perfectly matched to the head ports. 8A nozzle and .125 bypass jets are used in the injector to meter Mobil Premium gas.

A mallory Mini-Mag has been reworked by Charles Stockes and fires

through Crescent silicon primary wires. Autolite Standard electrode AT-1 plugs are set at .018.

Finally, the rear of the crank is fitted with a 12-pound Weber Flywheel, Hays 11-inch, 2800-pound clutch and Velvetouch disc.

And that's all there is to building a 200-mph, unblown, gas-fed Chevy. Sounds easy doesn't it? Jack claims it is, and his actions prove it!

Jack inserts the cam in the block with care. Just after this article was prepared, he cranked on new record at Bonneville at 204 mph. New car may even go nearly 250.

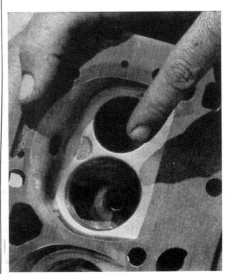
The combustion chambers were also polished, valves seated at their very edge.

If you are a Chevrolet racer—bracket, roundy-round, or otherwise—who could use the power potential of 450 or more cubic inches, but your racer can accommodate only the small-block engine because of its physical size, you may be able to have your cake and eat it, too, with a maxi-mouse powerplant, like those produced by Conely Speed & Machine Shop (807 East Grand River Avenue, Brighton, Michigan), where 454-inch small-blocks are commonplace and 480-inchers abound.

The guy who makes these monstrous motors happen is Jack Conely, a man who has had a close working relationship with Chevrolet Motor Division research and development programs since 1957. Jack built and drove brutal unlimited race cars on a Michigan circuit for many years, developing the big-inch small-block engines, first for his own use as injected nitro-burners in the unlimited car, and later figuring out combinations for alcohol, gasoline and carburetors. The things Jack learned on the big engines he has put to use in refining the construction of his NASCAR Grand National 355-inch engines for many of the biggest names in the Chevrolet camp, which is running away with the Winston series this year.

The heart of a Conely maxi-mouse —in 440, 454, or 482-inch form— is the cylinder block. The standard blocks simply won't work for these sleeveless engines, so Conely uses the limited production 0382318 or 3970010 Chevrolet blocks, the former a high-nickel-content alloy steel and the latter a high-tin-content alloy block. Both were developed for endurance racing applications by Chevrolet, based on their 400-inch passenger car block with its two center cylinders on each bank siamesed together with no water circulation between them. These blocks are less than ten pounds heavier than standard 350 passenger car blocks, but the increased weight is in all the right places and combines with the alloy properties themselves to afford incredible strength and durability, even in the 480-inch versions where walls are thinnest. The 440 version measures 4.1875 inches bore and 4.000 inches on stroke; the 454 measures 4.1875 inches in the bore and 4.125 inches on the stroke; and the giant 482-inch version measures 4.250 inches on both bore and stroke.

Conely and his staff of machinists and assemblers follow proven preparation techniques in doing the big-inch small-blocks with the following extra steps. All freeze plugs are replaced by new plugs, and the plugs are physically restrained from ever coming out by installation of cap-screwed crossbars. All blocks are relieved on the underside of the lifter bores to accept mushroom tappets and are thoroughly ground around the bottoms of all eight bores for rod cap rotating clearance. Lifter valleys are ground free of all casting flash and troughed out to aid drainback of oil. *All* oil drain holes are enlarged slightly in the valley area and then are fitted with epoxied-on mesh screens which will allow passage of oil but will prevent metal chips or valvetrain components from getting down inside the crankcase. On the deck surfaces of each block, Jack installs steel screw-in plugs in all water passages to create dry decks, mills the tops of the plugs off when the blocks are decked, and installs .062-inch copper O-rings in the finished decks, with the Chevrolet .017-inch stainless steel head gasket providing a trouble-free seal everywhere else. The upper plugs are then drilled out to maintain water circulation.

The crankshafts for all of Jack Conely's big-inch mouse motors are sourced by Moldex at about $600 a pop. The 4340 steel forgings are made with 2.450-inch main bearing journals, like the standard 350 engine, but have 2.200-inch journals for the rods, to accept the 6.135-inch big-block L88 rods (PN 3969804). The crankshafts use no center counterweights, but rather up to 15 slugs of Mallory heavy metal for balancing, since crankcase space is severely limited.

The rods themselves are crack-checked, beam-polished, lightened and shortened on the big ends to clear the block and the camshaft. Conely uses the best available Chevrolet 7/16-inch rod bolts with machined nuts to provide additional ro-

By Jim McCraw

BUILDING THE SMALL-BLOCK CHEVY INTO A FIRE-BREATHING MONSTER OF 440, 454 OR EVEN 482 CUBIC INCHES FOR ANY TYPE OF RACING YOU WANT TO DO

MAXI-MOUSE

ILLUSTRATION: DON MURRAY

tating clearance, but before the job is done, Jack routinely machines at least .060-inch of material from both rod and cap faces to shorten the whole assembly, touches up the beams and balancing pads with grinders, and then rebores the big end. The small end of the rod houses a full-floating, .990-inch 4340 steel wrist pin, locked in place by a .072-inch Spirolock ring. The short 2.920-inch pin is used, because in large-displacement engines, Conely uses only two rings: a 1/16-inch ductile moly Sealed Power top ring and a 3/16-inch oil ring that almost touches the pin in that portion of the land that passes *across* the pin hole.

The pin location for Conely's stroker small-blocks is set at .912-inch below the deck of the flat-top JE piston, and the thick-section but lightweight two-ringer pistons have held up beautifully in racing service despite the somewhat odd two-ring layout and high pin location with static compression ratios approaching 12.7:1. Not bad for a 475-gram piston, either.

An interesting sidelight to the 454 and 482 small-blocks is the meticulous boring and honing operation Conely uses for maximum ring and piston life in his engines. All operations are performed with a honing plate in place atop the block deck, and Jack uses a Sunnen 300 hone down to the last .005-inch. For the next .002-inch of metal removal, Conely uses a Sunnen AN301 stone. The next .002-inch is honed away with an AN500 stone, and the last .001-inch of material is taken off with an NN40-J85 or NN40-J87 stone set-up using manual Sunnen equipment as opposed to the automatic CK-10 Cylinder King machine. The process may sound like overkill and nit-picking, but if Conely didn't believe it contributes to the life of his engines, he wouldn't be doing it to each and every one of them.

With these kinds of displacement

MAXI-MOUSE

Conely maxi-mouse is housed in the 0382318 high-nickel alloy block, which is made like the 400 passenger car block, with two center cylinders on each bank siamesed, with no water in between bores for cooling. This block has stock-type fuel pump mount pad, full-width part number pad on right front of block, to resemble ordinary production pieces. Unfinished bore comes to 3.900 inches.

The high-nickel block is made with the 400-type siamesing, but it doesn't use the big 2.650-inch mains. Rather this racing piece uses the 2.450-inch main sizes from the 350 passenger car castings. Though it easily takes bores and strokes up to 4¼ inches without liners or sleeves, this block weighs only about 8 pounds more than the standard Chevrolet passenger car 350.

This big-inch block has been fitted with studs for head retention, and threaded socket-head plugs have been placed in the water passage entries to block them off, creating dry-deck block, using O-ring seals.

Valley of block on left has been ground in troughs and corners to aid oil return, with screens epoxied over drain holes to keep pieces out. Finished bore for 454 small-block is done in five-step cuts.

Freeze plugs in all of Conely's NASCAR 355 engines (as well as the monster motors) are standard plugs, but they are mechanically retained in the block by a cross-strap that's bolted into the block with capscrews.

figures and high compression ratios, Conely maxi-mouse engines don't need really radical cam and valvetrain components, just reliable ones. For that reason, Jack has worked exclusively with Bruce Crower for many seasons perfecting the equipment, including reduced-base-circle cams to clear the big ends of the rods, ranging in duration from 272 to 280 degrees at .050-inch lift; Crower rollers and 7/16-inch pushrods; Crower steel rocker arms in a variety of ratios; and Chevrolet valves, including a 2.060-inch intake and a 1.625-inch exhaust. It's all installed around Crower Vasco-Jet springs using 165 pounds of seat pressure with Crower and Chevrolet hardware and guide plates.

Standard cylinder head selection for the big-inch motors is the 340292 "turbo" casting from Chevrolet using 72cc combustion chambers. The heads have lots of hours of port grinding in them, based on the sum total of Jack Conely's experience with this type of small-block engine, which is considerable. All we can say is that the ports are enormous and very shiny from initial entry to final exit, and they are ported according to the fit of cast-rubber plugs that Jack keeps handy in the "flow room" of his unassuming but extremely busy shop. More detail than this on the cylinder heads we can't supply, except to say that Jack has run these engines using every type of experimental head Chevrolet has cast since 1963, including tunnel-ports, canted-valve aluminum, straight-valve aluminum, and several variations of the basic cast-iron "turbo" head; but he uses the standard 292 casting because it makes excellent horsepower and is widely available now. Besides standard head modifications, Conely drills and taps both ends of each cylinder head to accept a half-inch-pipe dry-sump scavenging pickup. Jack has a limited quantity of Chevrolet's

Continued on page 56

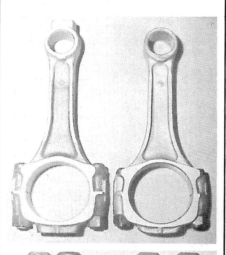

Conely prefers JE forgings for his big-inch mouse motors, and they are made to use 4340 .990-inch-diameter pins 2.920 inches long to fit behind the 3/16-inch oil ring. Piston array below shows 355-, 440-, 454- and 482-inch JE pistons designed by Conely.

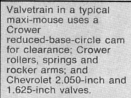

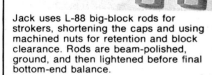

Jack uses L-88 big-block rods for strokers, shortening the caps and using machined nuts for retention and block clearance. Rods are beam-polished, ground, and then lightened before final bottom-end balance.

Valvetrain in a typical maxi-mouse uses a Crower reduced-base-circle cam for clearance; Crower rollers, springs and rocker arms; and Chevrolet 2.050-inch and 1.625-inch valves.

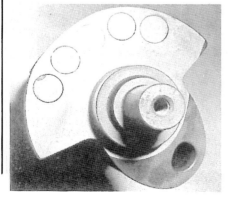

Moldex forgings are used for all Conely strokers, at $600 apiece, made with 2.450-inch mains and 2.200-inch throws for least friction problems. Each crank uses 15 slugs of Mallory heavy metal for balancing, with short pistons and big-block rods.

Torque Secrets Revealed

In a Street Engine, Torque is What You Want and This Small-Block Chevy Delivers.

By Dave Emanuel

PHOTOS BY COTTRELL RACING ENGINES

Whenever anyone talks (or should I say brags) about his or her street engine, invariably, the first question asked is, "How much horsepower does it make?" That's not surprising because horsepower is wonderful stuff. It's highly effective at impressing your friends, winning the favors of young maidens, and making top speed runs at Bonneville. But it's *torque* that moves a car from a dead stop and propels it down the street. What's more, even though horsepower is always in the spotlight, it owes its life to torque. Without torque there would be no horsepower.

Torque is twisting force as measured in pounds/feet. If an engine produces 400 lbs./ft. of torque, it exerts a twisting force that is the equivalent of a 400-pound block hanging from the end of a bar that is one foot in length. (In essence, torque is expressed in pounds per foot which is the reason that measurements are written as lbs./ft. rather than ft./lbs.—even though people say "foot pounds.")

(Editor's Note: the laws of physics—summation of moments about a point—suggest mathematically multiplying distance times load. This yields units of torque as ft./lbs. Frequently, automotive convention holds lbs./ft. to be the unit of torque. Actually, torque values remain unchanged with either style of units.)

A dynamometer measures torque. Horsepower is a computed (not measured) figure which relates torque to engine speed. So peak horsepower is actually a measure of an engine's ability to produce torque at high rpm. Typically, when an engine is modified to generate more horsepower, its ability to produce torque is not altered appreciably. What does change is the rpm level at which maximum torque is produced. With the formula for horsepower being Torque x rpm/5252, simply raising the point at which peak torque occurs can profoundly increase horsepower. At 5000 rpm, 400 lbs./ft. translates to 380 horsepower; at 3500 rpm, that same 400 lbs./ft. equals only 266 horsepower. While the 380-horsepower figure is much more impressive, in a street engine, 400 lbs./ft. of torque at 3500 rather than 5000 rpm is of considerably more use.

In selecting components for a high-performance street engine, which will spend a comparatively short portion of its life at or above 5000 rpm, the most sensible approach is to concentrate on peak torque, and let horsepower fall where it may.

Myron Cottrell of Cottrell Racing Engines reasoned that he could accomplish excellent results with a naturally aspirated 400-cubic-inch small-block capable of burning unleaded fuel. And

Measuring 406 cubic inches and producing 465 lbs./ft. of torque, this Myron Cottrell-built engine produces excellent torque due to the camshaft/cylinder head combination.

he figured he could achieve his goal at approximately 25 percent the cost of the twin-turbo system. Judging from the data produced by Cottrell on his Super-Flow dyno, he's been successful.

Cottrell began with a 2-bolt main 400-cubic-inch block which had been bored 0.030-inch oversize (yielding 406cid). After the block was cleaned and deburred, standard race/high-performance blueprinting and preparation techniques were applied, including boring and honing with torque plates installed, align-honing the mains, and cutting the deck. Cottrell notes that properly setting the deck height, also known as quench clearance, is of vital importance. By running close to zero deck, quench clearance works out to 0.035-inch to 0.040-inch with a standard composition-style head gasket. If quench clearance isn't set this tight, the engine is much more sensitive to fuel quality. However, care should be taken to retain acceptable compression ratios within the engine.

On the lower end, Cottrell employed a stock 400 crankshaft with radiused oil holes and proper journal finishing. For improved durability, the crank was Tufftrided. Rather than retaining the stock 400 connecting rods, 350 rods, which are slightly longer (and easier on the cylinder walls), were selected. Rod preparation included reconditioning, Magnafluxing, installation of new bolts, and an alignment check. These were mated to Sealed Power forged flattop pistons (PN 7066P) surrounded by Total Seal Gapless® rings.

To keep costs in line and to avoid problems with availability, Cottrell selected large-chamber 441 "smog mo-

In addition to machining the block decks, Cottrell chamfered the head bolt holes, and cleaned the threads with a tap.

(left) Gapless® piston rings from Total Seal were file-fitted to each cylinder and checked for proper gap.

Camshaft profile is vital to maximum torque. To enhance performance even further, selection of valvetrain componentry was made with the goal of reducing friction and increasing stability. Cam Dynamics roller rockers were used in conjunction with $^{7}/_{16}$-inch-diameter studs, heavy-duty pushrods, a roller timing chain, and stock-type valvesprings and retainers. Valve diameters measure 2.055-inch intake and 1.60-inch exhaust.

THE TOOLS OF TORQUE

Coaxing a small-block Chevrolet to produce one lb./ft. of torque per cubic inch of displacement is an easy task. All you have to do is apply a few well-proven engine modification techniques and you're there. Moving substantially beyond that figure is quite another matter. In the case of the Cottrell-built 406-cubic-inch engine, the 465 lbs./ft. peak torque reading is the result of the engine builder keeping his eye focused on his original objective.

If *maximum torque*—as opposed to maximum horsepower—is the goal, camshaft duration must be relatively short. In this instance, duration at 0.050-inch lift is approximately 220 degrees on the intake side and 230 on the exhaust. Lobe separation was also selected to maximize torque. Lobe center (not to be confused with lobe centerline) is the angle between the centerline of corresponding intake and exhaust lobes. When building an engine for torque, remember that a smaller angle will increase overlap and add usable low-speed torque. Increasing the angle will shift the torque curve towards the upper rpm range. As opposed to most off-the-shelf cams, which are typically designed to increase horsepower at the upper end of the rpm scale, Cottrell's camshaft grind produces peak torque at low engine speeds. And by design, some top-end horsepower was compromised to achieve the goal. Note that peak horsepower (387) is achieved at 5250 rpm, and then power falls off sharply by 30 horses at 5500. Valve timing and port size are simply too restrictive for operation in the 5500-plus rpm range.

Given the cam's duration, maximum valve lift is in the 0.450-inch to 0.460-inch range. (Higher lifts with that amount of duration will lead to a high rate of wear and possibly, premature cam failure.) That being the case, the cylinder heads had to be modified with a goal of maximizing low-lift flow. Installation of 2.05-inch intake valves may seem to contradict this, but actually makes good sense. The resulting reshaping of the short turn radius improves low-lift flow, and the larger diameter helps to compensate for the low lift. No work was done to increase air flow at valve lifts above 0.450-inch because the valves are never raised that high.

While the camshaft is certainly a major influence on an engine's torque-producing capability, cylinder head modifications also play an important role. By having the heads tailored for maximum low-lift flow, Cottrell was able to augment the effect of the cam by improving cylinder filling. Had the ports been made larger, or time been spent improving high-lift flow numbers, peak horsepower would have increased, but low-speed torque would have dropped. In this instance, that would have been counterproductive. Additionally, torque was helped up the scale by the Performer dual-plane intake manifold and the Quadrajet carburetor with its vacuum-actuated secondaries. All the tools of torque were put to excellent use.

JUNIOR TORQUE MONSTER

After completing the 406 engine project, Cottrell started wondering about applying the same technology to a 350. With a slightly different selection of equipment, as requested by one of his customers, he built a 350 small-block and found that while it produced excellent torque readings, it did so at a somewhat higher rpm level. Naturally, the horsepower was also increased. With the 350, peak torque (400 lbs./ft.) was recorded at 4750 rpm, maximum horsepower (412) was achieved at 5750 rpm, and torque exceeded 350 lbs./ft., from 3000 to 5750 rpm.

Considering the smaller displacement, the lower torque reading (as compared to the 406) is expected, but Cottrell figures that camshaft duration was too long, hence the 4750 rpm torque peak. Future plans call for additional testing with a slightly shorter duration cam, which should drop the peak torque point down to the 3500-4000 rpm range. These cams, as well as cam and cylinder head assemblies, are available separately for car crafters desiring to "torqueize" existing small-block engines.

Cottrell used 5.7-inch connecting rods and Sealed Power pistons, requiring some slight modification to the crankshaft.

tor" cylinder heads. He states, "These castings are readily available and not as prone to cracking as the newer ones. But for extra insurance, we Magnaflux all the castings before we use them. These heads have combustion chambers that have about 72 cc's, which allows us to keep the quench area tight and still use a flattop piston for the best burn. The result is a final compression ratio of about 9.5:1 and an engine that is not octane-sensitive."

Cottrell's head porter took a highly specialized approach to cylinder head preparation. In conjunction with a mild porting job, he installed stainless steel 2.055-inch intake and 1.60-inch exhaust valves. These particular diameters were used not only for their size and flow potential, but also because of

CARE FOR A LITTLE INTERNAL COMBUSTION?

We're pretty fired up about our new line of spicy sauces and salsas. One taste and you will be too. Created for Hot Rod Magazine by Dave's Gourmet, these "Combustion Concoctions" come in two flavors, each with its own collector's edition label. Try both varieties—the 1934 3-window Coupe and 1927 Track Roadster. To order, call 800.758.0372.

HOT ROD MAGAZINE

HOT SAUCE AND SALSA

The press-in stud bosses must be milled with a spot facer, as shown, *before* tapping the existing hole.

As opposed to horsepower-directed engine, torque engine can use many standard over-the-counter high-performance parts with little or no modification. The Sealed Power pistons used ring lands of 5/64-inch and 3/16-inch, the standard of the industry.

Running minimal quench clearance is one of the keys to building an engine that is relatively insensitive to fuel octane. Large chamber heads were used so that flattop pistons installed with zero deck would provide a compression ratio of approximately 9.5:1. Piston-to-cylinder wall clearance was set at 0.0035- to 0.004-inch.

the treatment that can be given to the short turn radius. After cutting the seats to accept these valves, (especially the intakes) a smooth, well-shaped short turn radius can be developed that is extremely effective at lifts up to 0.460-inch. Cottrell emphasizes that this is an important concept because if the engine is to produce the desired performance profile, a relatively mild camshaft must be selected. It is therefore essential that the flow characteristics of the head be tailored to be compatible with cam lift.

Camshaft selection is important to net the highest torque numbers. The camshaft used here is a hydraulic dual-pattern grind with a gross lift of approximately 0.450-inch and durations (at 0.050-inch lift) of 221-degree intake and 230-degree exhaust. The camshaft is cut on a 112-degree lobe centerline. The cam is driven by a roller timing chain and operates the valves through 1.5:1 roller tip rockers mounted on premium-grade 7/16-inch studs. (Cottrell has found that 1.6:1 rockers offer no benefit with this combination.) Spring pressure is set at a mere 110 pounds on the seat (with the engine designed for a 5500 rpm maximum, super-stiff springs are not required). This combination of equipment was selected as a means of reducing friction while offering maximum valvetrain rigidity.

Cottrell selected stainless steel intake and exhaust valves which are swirl-polished to aid flow, in order to make the engine compatible with unleaded fuels. With low-rpm profile of the engine, valvespring pressure was set at 110 pounds closed.

DYNO TEST RESULTS

RPM	Corrected Torque (lbs./ft.)	Corrected HP
2250	390	167
2500	402	191
2750	437	229
3000	463	265
3250	**465**	288
3500	**465**	310
3750	463	331
4000	454	346
4250	448	364
4500	439	376
4750	422	382
5000	401	382
5250	387	**387**
5500	341	357

Dyno test conducted by Myron Cottrell on a SuperFlow 901 dyno. Test method was steady-state mode, starting at maximum rpm. Load was increased progressively to reduce rpm for each successive test point.

Manifold vacuum at idle:
17 in./Hg at 850 rpm in neutral
14.5 to 15 in./Hg in gear

TORQUE SECRETS

After the engine was assembled, Cottrell topped it off with an Edelbrock Performer and a Quadrajet carburetor. Again, this combination of equipment was selected because of its reasonable cost, easy availability, and dual-plane design's compatibility with a torque-oriented engine combination. Cottrell also did some testing with a Holley 750 cfm double-pumper in place of the Quadrajet, but found that the Holley's small torque increase (less than 5 lbs./ft.) was offset by a reduction in fuel economy.

As can be seen from the accompanying chart, the maximum horsepower number that Cottrell produced in his dyno testing (387) is stout, yet not extraordinary, considering it was extracted from 406 cubic inches. But look at the torque figures—465 lbs./ft. at 3500 rpm, 402 lbs./ft. at 2500 rpm and 401 lbs./ft. at 5000 rpm. Those are big-block numbers. And talk about a flat torque curve! From 2500 to 5000 rpm, torque is above 400 lbs./ft. If that's not the torque of the town, then nothing is.

Camshaft was carefully degreed before final installation. Cottrell starts by accurately establishing top-dead-center.

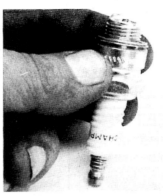

According to Cottrell, cutting back the side of the spark plug electrode is worth five to eight horsepower. In his super-torque street engines, Cottrell runs Champion J-6JC spark plugs gapped at 0.040-inch. The zap is provided by a MSD-6 module.

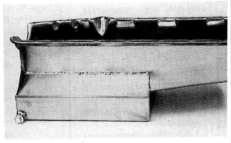

Labeled with the Champ Pans moniker, the JR racing oil pan features excellent baffling for high-performance street use.

SOURCES

Cottrell Racing Engines
Dept. CC
4255 County Road 10 East
Chaska, MN 55318
612/448-5330

Edelbrock Corp.
Dept. CC
411 Coral Circle
El Segundo, CA 90245
213/322-7310

Speed-Pro/Sealed Power Corp.
Dept. CC
100 Terrace Plaza
Muskegon, MI 49443
616/724-5011

Total Seal Piston Rings
Dept. CC
2225 W. Mountain View, Suite 17
Phoenix, AZ 85021
602/242-9421

MAXI-MOUSE

Continued from page 51

3824059 aluminum water pumps which he uses on the big engines as a further weight saving, since they pump about the same as the standard cast-iron water pumps.

To light off the highly compressed mixture in those tiny chambers, Conely uses the complete Delco electronic ignition system, PN 3997782, which includes tach-drive distributor, control box and wiring harness. To get the message to the plugs, Conely uses 7mm solid-silicone-insulated stainless steel-wire ignition cables, which he will soon begin to sell in the aftermarket under his own brand name. Jack has tried to destroy this wire with heat, chemicals, abrasives, corrosives and solvents and has found it unkillable, so he'll be marketing it soon with molded straight and 90-degree ends in sets.

To exhaust the chambers, Jack recommends that his stroker customers use 1⅞-inch headers with 32-inch-long primaries and collectors from 14 inches long and longer, depending of course on the type of car and its application. We know it sounds strange, but those are Conely's considered opinions for his big-inch engines.

And if you're not using nitromethane and/or alcohol with Crower port injectors, which Jack considers a standard setup for many of his roundy-round customers, he will fit the engine instead with an 830 Holley double-pumper, a Holley Strip Dominator small-block manifold, and a two-inch-thick cast-aluminum spacer plate. This combination will yield in the neighborhood of 600 horsepower on good gasoline, lubricated and cooled as they all are when Conely builds them by a custom-made oil pan and Weaver dry-sump pump and tank system for competition applications. Of course such a 440, 454 or 482 could also be lubed by the more conventional wet-sump, deep-pan and heavy-duty pump system if the customer was interested in a big-inch small-block for street, drag or bracket racing use. The advantages of a lightweight, compact powerplant that takes up small-block space and puts out big-block power are not hard to see, and the life-giving tricks that Jack Conely has devised after years of building, tuning and racing these maxi-mouse engines make them as nearly bulletproof as a combination like this can be. GM may be phasing out the big Mark engines, but the small-block will be with us for at least a few more years, and Jack Conely will be building the big ones as long as there is a demand for them. **HR**

Reality Mouse

By Maximizing Cubic Inches, Installing Efficient Cylinder Heads, And Using Bolt-On Parts, Can A Small-Block Chevy Make 450 Horsepower On 92 Octane?

POWER COMBOS — MAX HP ON PUMP GAS!

By Ed Taylor

IN A PERFECT WORLD, horsepower would be king, fuel would be cheap, and lead would be nontoxic. But, the reality is making mega-power on today's pump gas just won't happen if you are still using the conventional tricks taught as gospel in the days when musclecars roamed freely.

Sure, fuel octane ratings of 100-plus can be purchased for a handsome price, but you better bring your Visa card.

Making big horsepower with today's unleaded fuels can be accomplished, but it takes more than just a torque wrench to build such a combination. To know where you're going, you need to set some priorities. With the utopian desires of CC Editor Chuck Schifsky and Tech Editor Marlan Davis, a few parameters were set: (1) the engine must idle below 1000 rpm (that eliminated the use of high-lift, long-duration camshafts with lobe centers wider than the expanse of several ex-CAR CRAFTERs' backsides); (2) the engine must produce more than 14 in/Hg of manifold vacuum to ensure that vacuum-assisted accessories, including power brakes, would function properly; and (3) the package must run on premium unleaded pump gas (in our case, 92 octane). Our goal was 450 hp.

With the parameters set, we enlisted the advice and efforts of engine builder Ken Duttweiler to help us in our quest to find pump gas power. After he shook his head and raised a single eyebrow, one of Ken's first suggestions was to add a fourth parameter—build the engine using as many off-the-shelf parts as possible to help keep the cost down, thus making it more attractive to the average Joe. In other words, no NASCAR engines.

With that in mind, Duttweiler suggested that we go for a 406cid Chevy

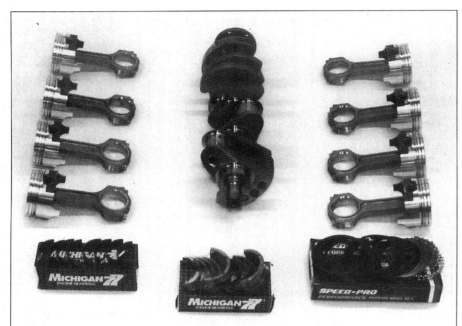

The reciprocating assembly for the 406 includes Summit Engine Shop's Pro Crank for the 406, Summit 5.7-inch I-beam rods, KB pistons, and Clevite 77 bearings. The reduced weight of the assembly allows us to rev a little higher and seek the upper limits of the rpm band. The Summit Engine Shop's Pro Crank is a forged 4130 steel alloy component that offers a price competitive with a over-the-counter cast crank—$599. With a weight nearly 10 pounds less than its OEM equivalent, that equals quicker revs. The generous rod fillet requires the use of high-performance series Clevite 77 rod and main bearings. Note the large chamfer for fillet clearance on the bearing shown above.

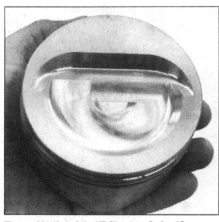

The combination of the KB Signature Series 18cc dished piston along with TFS' 64cc Twisted Wedge cylinder heads resulted in a compression ratio of 10.25:1. While that figure is not necessarily the moderate compression pump gas engines require, the spark plug placement with the TFS heads offer excellent flame propagation thanks to the spark centering in the cylinder bore. Less timing lead is required for the spark to ignite the air/fuel mixture across the entire face of the piston.

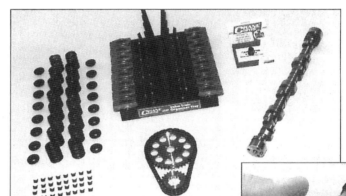

The complete valvetrain for the Summit 406 is from Crane and includes valvesprings, retainers, keepers, rocker arms, pushrods, solid roller-tappet lifters, cam button, camshaft, and 8-degree advance/retard billet timing set.

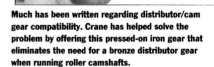

Much has been written regarding distributor/cam gear compatibility. Crane has helped solve the problem by offering this pressed-on iron gear that eliminates the need for a bronze distributor gear when running roller camshafts.

combination. This Mouse is built without the expense of an offset-ground stroker crank because it uses a 400 crank in a 0.030-inch-over 400 block. While we could have built a smaller engine, we wanted to achieve a good peak horsepower figure without sacrificing low-speed torque. Should we have gone the 350-cubic-inch-or-smaller route, this would have required the use of a higher stall-speed torque converter or a higher ring-and-pinion ratio—both items that would have affected street drivability and economic use.

With Ken's suggestions in mind, we began our search for a "seasoned" 400 block from our local salvage yard where we found a two-bolt-main block that had served its former station wagon well. We loaded it up and took it to Chris Skercevic at Skercevic Automotive Machining in Ventura, California, where it was hot-tanked, sonic-checked, and Magnafluxed to find any flaws.

While Chris began the machining process, we turned to Summit Racing Equipment and ordered the rotating assembly. The feature part is Summit's Pro Crank; an internally balanced, forged-steel crank that is offered at a price that beats the expense of purchasing a stock cast-iron crank, and then having it inspected, properly prepped, and balanced. Accompanying the $599 crankshaft was a set of Summit 5.7-inch I-beam connecting rods, Clevite 77 high-performance series main and rod bearings, and Keith Black Signature Series dish-top pistons that Duttweiler fitted with a set of Speed-Pro file-to-fit rings with the end gap set between 0.020 and 0.025 inch. When the 18cc dish-top pistons were combined with 64cc combustion chambers we achieved a net compression ratio of 10.25:1 based on Duttweiler's computations. In addition, the reciprocating assembly is noticeably lighter than the stock 400, a fact that would allow this engine to rev higher. Rounding out the internally balanced reciprocating assembly is a Summit-supplied harmonic balancer and flexplate. The whole package, including balancing, was less than $2000.

For the oiling system, Summit recommended a Moroso 7-quart oil pan and pickup used in conjunction with Moroso's Oil Control Kit. This windage tray assembly includes a crank scraper and tray that prevents power-robbing oil splash from slowing down the crank.

Camshaft selection was made simple as we contacted Crane Cams' tech line and explained the parameters of our engine

REALITY MOUSE

The cam sprocket of the Crane billet timing set offers 8 degrees of advance and retard adjustment in 2-degree increments.

On the left is the larger 1.460-inch-diameter Crane valvespring that replaced the smaller spring that comes standard on the TFS Twisted Wedge cylinder heads. This is necessary due to the use of the roller cam, which requires higher spring pressure at the seat.

Parts List

Description	Manufacturer/Supplier	Part Number
Cam button	Crane Cams	PN 99164-1
Camshaft (solid roller)	Crane Cams	PN 118511
Carburetor (750cfm mech. secondary)	Holley	PN 0-4779
Connecting rods (5.7 inch)	Summit Racing Equipment	PN SES-3-60-05B-570
Crankshaft	Summit Racing Equipment	PN SES-3-49-05-C57
Cylinder heads (Twisted Wedge)	Trick Flow Specialties	PN TFS-31400001
Distributor (billet)	MSD	PN 8555
Distributor cap	MSD	PN 8437
Distributor rotor	MSD	PN 8462
Engine dress-up kit	Automotive Racing Products	PN 534-9801
Flexplate	Summit Racing Equipment	PN SUM-G-100SFI
Fuel pump	Edelbrock	PN 1721
Fuel pump pushrod	Moroso	PN 65750
Gasket set	Fel-Pro	PN 2802
Harmonic balancer	Summit Racing Equipment	PN BMM-64262
Harmonic balancer bolt	Automotive Racing Products	PN 234-2501
Head bolts	Automotive Racing Products	PN 134-360
Head gaskets	Fel-Pro	PN 1010
Ignition box (6AL)	MSD	PN 8420
Intake manifold—Victor Jr.	Edelbrock	PN 2975
Lifters	Crane Cams	PN 111519-16
Main studs (w/windage tray)	Automotive Racing Products	PN 234-5501
Oil control kit	Moroso	PN 23035
Oil pan (7 quart)	Moroso	PN 20190
Oil pump driveshaft	Moroso	PN 22070
Oil pump, high-volume/pickup kit	Summit Racing Equipment	PN SES-3-60-08-024
Pistons	Keith Black/Summit	PN UEM-KB147-030
Plug wire set	MSD	PN 3165
Pushrods (5/16x0.100+)	Crane Cams	PN 11632-16
Quick jet change bowl kit	Holley	PN 34-24
Retainers	Crane Cams	PN 99953-16
Rocker arms (alum. extruded roller)	Crane Cams	PN 11755-16
Starter	Summit Racing Equipment	PN SUM-G-1660
Timing chain set (billet)	Crane Cams	PN 11984-1
Timing cover	Summit Racing Equipment	PN SUM-G-6300
Valve covers	Trick Flow Specialties	PN TFS-31400902
Valve keepers	Crane Cams	PN 99097-1
Valvesprings	Crane Cams	PN 99893-16
Water pump	Edelbrock	PN 8810

and the test. The cam recommended was Crane's Fireball Roller (PN 118511). This cam specs out at 236 degrees intake duration and 244 degrees exhaust at 0.050-inch lift and 0.525 inch intake and 0.543 inch exhaust of gross valve lift with a 112-degree lobe separation. With the cam we also requested the appropriate cam kit for assembly at Duttweiler's.

With the pulse of the engine set by Crane, it was now time to determine respiration. Keeping in mind both performance and price, we sought to acquire one of the first sets of Trick Flow Specialties (TFS) Twisted Wedge Heads for the small-block Chevy. These aluminum heads retail for $995 assembled and are ready to bolt on your Mouse engine including the swirl-polished 2.02/1.60 stainless valves. In addition, they will accept all stock valvetrain components, yet feature relocated valve angles that unshroud the intake valves, a new spark plug location closer to the bore centerline, a double-squish quench area, and a port shape with adequate volume to promote airflow. But how would these budget heads perform with the demands of an even higher-lift, long-duration, roller-lifter camshaft than we had chosen? A few minor changes are needed in the plan to make this truly a pump gas powerplant.

These changes include upgrading the TFS-supplied valvesprings with larger-diameter double springs from Crane. In addition, our camshaft was of the small base-circle variety, requiring pushrods 0.100-inch-longer than stock. We also changed over to 7/16-inch rocker studs to offer a more stable valvetrain and eliminate the need for a stud girdle.

The induction system for the Reality

Moroso's Oil Control Kit is designed to work with the 7-quart Moroso pan to keep power-robbing oil splash off of the rotating parts.

Here, the windage tray is being mounted using extended-length ARP main studs. After installation on the studs, clearance between the tray and the rotating assembly must be checked.

Providing spark for the 406 was a complete system from MSD including the MSD-6AL, Blaster II coil, Pro Choice billet distributor, and 8mm Heli-Core plug wires.

Holley Quick Change float bowls were installed to facilitate jet changes during the dyno evaluations. This kit is available from Summit or your local speed shop that carries Holley accessories. It's well worth the investment.

Mouse was accomplished with an out-of-the-box 750cfm Holley 0-4779 double-pumper atop an Edelbrock Victor Jr. intake. Edelbrock also supplied the aluminum water pump, while spark came via a MSD-6AL ignition. Fel-Pro donated the engine gasket set. Now we had all the components needed to get our concept on Duttweiler's dyno.

After carefully measuring all the parts, Duttweiler began the assembly process, adding a little margin to the clearances so we could run the 406 a tad harder and faster then you would your new street engine. Once Duttweiler had finished the assembly, the small-block was bolted to the dyno, then a set of 1⅝-inch Hooker headers routing emission to a pair of Borla XR-1 mufflers was installed—now the testing could begin. After proper break-in, we adjusted the valve lash and dialed in the baseline setup. A pair of jet changes and several experimental pulls helped to determine the amount of initial timing. Immediately, the results were noteworthy—460 horsepower at 5600 rpm and 475 lb-ft of torque at 4400 rpm. This was accomplished using #70 and #78 primary and secondary jets, respectively, and a total of 32 degrees of advance. Even more impressive was the small-block's low-rpm torque, which measured 438 lb-ft of torque at 2600 rpm.

GOOD CHEVY READING
FROM
CarTech

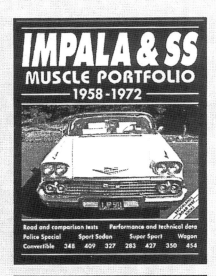
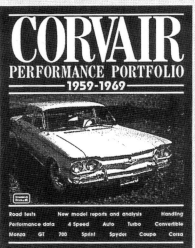
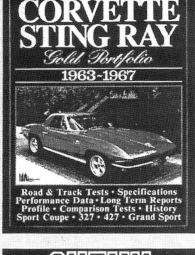

REALITY MOUSE

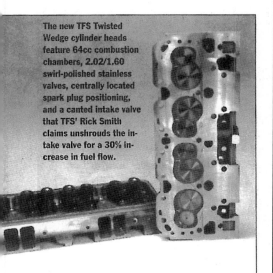

The new TFS Twisted Wedge cylinder heads feature 64cc combustion chambers, 2.02/1.60 swirl-polished stainless valves, centrally located spark plug positioning, and a canted intake valve that TFS' Rick Smith claims unshrouds the intake valve for a 30% increase in fuel flow.

Test Results

RPM	Test 1: Baseline		Test 2: 1¾-inch Hooker Headers		Test 3: 1.6:1 Crane Rocker Arms	
	Torque (lb-ft)	Power (Hp)	Torque (lb-ft)	Power (Hp)	Torque (lb-ft)	Power (Hp)
2600	439.5	220.5	435.1	215.4	423.6	209.8
2800	445.5	237.5	438.9	234	434	231.4
3000	444.8	254	447.2	255.5	448.5	256.2
3200	444.6	270.9	451.7	275.2	459.4	279.9
3400	443.1	286.8	451.6	292.4	459	297
3600	449	307.8	449.9	308.3	456.9	313.2
3800	460	332.9	447.1	323.5	453.5	328.1
4000	465.1	354.2	442.3	336.9	450.3	342.9
4200	473.4	378.5	451.7	361.2	456.4	365
4400	**474.6**	397.6	462.4	387.4	465.2	389.7
4600	472.2	413.6	469.7	411.4	470.7	412.2
4800	470.5	430	470.5	430	**474.2**	433.3
5000	465.2	442.8	470.3	447.7	470.8	448.3
5200	458.1	453.5	465.5	460.8	468.6	463.9
5400	445.9	458.5	456.1	468.9	461.8	474.7
5600	430.4	**459**	444.5	474	450.1	480
5800	413.8	456.9	431.3	476.3	436.2	481.8
6000	396.2	452.6	417.1	**476.4**	424.5	**485.1**

*All figures corrected to 28.82 in/Hg. Peak values in **bold** type.*

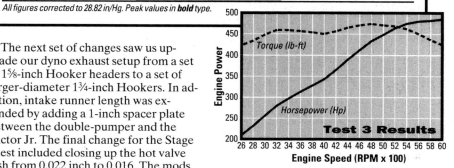

After replacing the 1.5:1 rocker arms with 1.6:1 Crane roller rockers in our search for more power, we relashed the valves at 0.018 inch. This was done to maintain the lash spec used with the 1.5:1 rockers.

The next set of changes saw us upgrade our dyno exhaust setup from a set of 1⅝-inch Hooker headers to a set of larger-diameter 1¾-inch Hookers. In addition, intake runner length was extended by adding a 1-inch spacer plate between the double-pumper and the Victor Jr. The final change for the Stage 2 test included closing up the hot valve lash from 0.022 inch to 0.016. The mods paid off as power improved 15 horsepower with 476 peak horsepower at 5600 rpm and 470 lb-ft of torque at 4600 rpm. Low speed torque showed a negligible decline to 435 lb-ft at 2600 rpm.

For the final dyno session, we swapped the 1.5:1 rocker arms for 1.6:1 Crane roller rockers, increased carb jet size to 72 primary/80 secondary, and upped the timing to 34 degrees total. Duttweiler's suggestions kept us going in the right direction as we improved to 485 hp at 5900 rpm and 475 lb-ft of torque at 4800 rpm. As could be expected, low speed torque fell to 424 lb-ft, still enough to easily fry the hides on your favorite street terror.

All things considered, we are very pleased with the 406 package, especially the TFS Twisted Wedge Heads. While these heads were impressive in our application, we think they'll work just as well as stock replacements. Sure, we could have spent more money, invested more time, and made more horsepower, but the results would have been a thinner wallet and an engine whose dual purpose may have been compromised. Reality is what you live with, and what more could you ask for from a total investment of less than $4500? **CC**

Sources

A.E. Clevite Engine Parts
Dept. CC
325 E. Eisenhower Pkwy.
Ann Arbor, MI 48108-3388
313/663-6400

Automotive Racing Products (ARP)
Dept. CC
250 Quail Court
Santa Paula, CA 93060
805/525-5152
800/826-3045

Borla Performance Industries, Inc.
Dept. CC
5901 Edison Drive
Oxnard, CA 93033
800/927-5129

Crane Cams
Dept. CC
530 Fentress Blvd.
Daytona Beach, FL 32114
904/252-1151

Duttweiler Performance
Dept. CC
1563 Los Angeles Ave.
Saticoy, CA 93004
805/659-3648

Edelbrock Corporation
Dept. CC
2700 California St.
Torrance, CA 90503
310/781-2222

Fel-Pro Inc.
Dept. CC
7450 N. McCormick Blvd.
Skokie, IL 60076
708/674-7700

Holley Replacement Parts
Dept. CC
601 Space Park N.
Goodlettsville, TN 37072
615/859-3124

Hooker Industries, Inc.
Dept. CC
1024 W. Brooks St.
Ontario, CA 91762
909/983-5871

Moroso Performance Products, Inc.
Dept. CC
80 Carter Drive
Guilford, CT 06437
203/453-6571

MSD/Autotronic Controls Corp.
Dept. CC
1490 Henry Brennan Drive
El Paso, TX 79936
915/857-5200

Speed-Pro Sealed Power Replacement
Dept. CC
100 Terrace Plaza
Muskegon, MI 49443-0299
616/724-5200

Summit Racing Equipment
Dept. CC
P.O. Box 909
Akron, OH 44309-0909
216/630-0240 (tech)
216/630-0200 (orders)

Trick Flow Specialties (TFS)
Dept. CC
1248 Southeast Ave.
Tallmadge, OH 44278
216/630-1555

STRIP-FLOGGING PAW'S 383 STREET SMALL-BLOCK

By Mike Johnson

There is a changing of the guard occurring on the streets these days. The evolution of the small-block Chevy has gradually assumed increasingly larger proportions. In the '50s, the small-block Chevy was exemplified by the 283. By the '60s, the Mouse motor had matured to 327 inches, and in the '70s it bulked up into the 350. But now the savvy street 350 has inherited the long arm of its 400-inch brother and has been transformed into the larger mini-Rat 383.

What has fostered this move to ever-larger Mouse mutations? Simple: Cubic inches rule the streets, and what easier way to build a better Mouse than to just build it *bigger*? A 383 is actually nothing more than a .030-inch overbored 350 stuffed with a turned-down 400 crankshaft. The end result is a relatively inexpensive Mouse motor approaching Rat proportions, which means added low-speed power for the street.

Certainly, 383 small-blocks are nothing new. Performance Automotive Wholesale (PAW), among many others, has been selling these engine kits for years. But we thought it would be fun to perform a real-world evaluation by bolting one of these 383-inch engine packages in a street car and flog it in a number of different configurations. Our goal? Stuff this hummer into an early Chevelle and trip the lights to the tune of high 12s without sacrificing street manners.

Our test subject was a 3450-pound '65 SS Chevelle with a *mucho*-tired 350, outfitted with a very strange combination of an Edelbrock S.P.2-P intake, an Economaster 400-cfm carburetor, stock exhaust manifolds, dual exhaust with glasspack mufflers, a Turbo 350 trans, and 3.08 rear gears in an open rearend. Our baseline trek to Los Angeles County Raceway (LACR) netted a less-than-outstanding corrected time of 15.87/86.23 on street tires and a 15.52/87.28 with a set of Mickey Thompson Indy Profile S/S G60-15 tires. This motor was obviously tired, as evidenced by the weak mph run that roughly measured out to be about 180 stormin' ponies. It was obviously time for an engine swap.

We obtained one of PAW's basic 383-cid engine kits and brought it back to the garage for an afternoon assembly session. The kit consists of a four-bolt main 350 block, fully machined, with the cam bearings installed along with a machined 400-cid crankshaft to bolt into the 350 block. The TRW L2403 .030-over forged aluminum pistons were already mated to resized 400-cid rods fitted with high-quality rod bolts. The rings are Speed-Pro single-plasma moly with a cast-iron second ring and stainless-steel oil-control rings. The only things we had to add were a 400-cid harmonic balancer and flexplate.

To begin this project, we decided to outfit the engine with a stock, 350-type hydraulic camshaft, 1.94/1.50-inch valve heads, and a stock intake and exhaust system. This way, we could establish a baseline with a *verrry* stock engine to see the performance improvements between the stock engine and our add-on goodies. In addition to the new engine, we also installed a fresh B&M Turbo 350, slightly higher stall speed 11-inch B&M torque converter, B&M Quicksilver shifter, and trans cooler. The transmission, converter, and cooler installation were made mainly for durability reasons because the Chevelle's 350 trans and converter were unknowns and we didn't want to be forced to rebuild transmissions midway through the test.

Not surprisingly, the combination of the stock but fresh 383 and slightly better converter picked the car up by a full .62 second to a 14.90/90.33-mph pass, with the car spinning the tires part way through first gear and shifting at a low 4500 rpm. Our game plan then called for a change of gear ratios, so we chose a set of Zoom 3.55 gears to replace the 3.08 ratio and managed to scrounge up a three-series posi carrier (which is no small task since these pieces are extremely rare). With the gears installed we again boogied up to LACR, where the Chevelle picked up an additional .14 second to a 14.76/90.53, which was almost exactly what our Quarter Jr. computer program predicted.

While our mini-Rat significantly improved the low-speed power, it fell off dramatically at rpm higher than 4500, which we theorized to be a severe exhaust restriction. This led us to our next phase of the performance plan, which was a set of Hooker 1¾-inch headers, a pair of Hooker's latest Super Competition turbo mufflers, and a complete 2¼-inch dual-exhaust system with a balance pipe installed by Champion Muffler in Simi Valley, California. On the induction side, we added an Edelbrock Performer dual-plane intake, 750-cfm 3310-2 vacuum secondary carburetor, and a complete Mallory Unilite electronic ignition system with Pro wires.

With all the pieces bolted in place, we once again made the trek to the track

The basic PAW 38 long-block kit com with the major cor nents you see her with TRW forged dished pistons for compression, Speed-Pro moly rings, camshaft of your choice, a double-ro chain, and completely machined bloc heads (with 1.94/1.50-inch valves), cr and rods ready for assembly.

MAXI-MOUSE

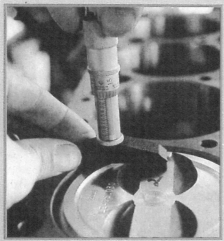

We carefully checked all clearances as we assembled the PAW 383, including the ring end gap, all rod and main bearings, crank endplay, rod side clearance, piston deck height, and even the volumes of the piston dish and combustion chamber. The motor computed to a very close 8.98:1.

only to be completely surprised by the results. Despite the significant bolt-ons that we added, the Chevelle picked up absolutely no additional e.t. or mph. Zippo! Jetting and ignition changes made insignificant improvements in the Chevelle's e.t. or mph. We noticed, however, that the shift point was still 4500 rpm, and that revving the engine higher only resulted in slower e.t.'s. It didn't take us long to determine that the 383 was seriously camshaft limited with its stock 350 hydraulic cam (.390/.410 lift with 195/202 degrees of duration at .050-inch tappet lift).

This led to our next phase: a camshaft swap. We had originally planned to bolt in a Crane HMV-272 hydraulic cam, but after reviewing the specs, decided instead on one step larger with the HMV-278. The installation process was handled easily in an afternoon and only required removing the radiator and jacking the motor slightly. We degreed the cam in "straight up" without advance or retard and double-checked the figures just to make sure. We also broke in the cam properly by varying the rpm between 1600 and 2400 rpm constantly for the first 20 minutes of operation to ensure proper lubrication of the new cam and lifters.

There was an instant improvement in all-around power with the addition of the new cam without the expected loss in low-speed torque. The engine also had a nice lopey idle, which sounded great at the stoplights. The first shot down the dragstrip confirmed our seat-of-the-pants evaluation with a stout series of 14.0s at 97.11 mph. With a few jetting and timing excercises, and the addition of a set of Crane 1.5 roller rocker arms,

MAXI-MOUSE

We started with a stock 350 hydraulic cam (above) for our initial test. Later, we evaluated the difference in bolt-ons—including a Crane HMV-278, which made a tremendous difference in the Chevelle's performance.

Induction and ignition pieces (right) included an Edelbrock 2101 Performer dual-plane intake, a Holley 3310-2 750-cfm carburetor (which we later converted to a secondary metering block), a Carter mechanical fuel pump, and a Mallory Unilite distributor with matching coil and wires.

For drivetrain durability we removed the stock Turbo 350 trans and installed a B&M trans, B&M 2500-rpm-stall 11-inch converter, a B&M Quicksilver ratchet shifter that fit nicely into the Chevelle's automatic console, and a heavy-duty B&M trans cooler.

After the baseline test of the new 383 engine, we also bolted in a set of Zoom 3.55 gears along with a posi unit to replace the stock 3.08 gears. The net improvement was almost .15 second, but it did put the motor closer to its power range through the lights. The gear is also completely streetable, cranking 3000 rpm at 65 mph.

we managed to put the Chevelle firmly into the 13s with a corrected 13.79/99.44-mph run, just shy of a full second quicker than the previous 14.74 e.t. and almost nine mph faster!

While the camshaft at first may appear to be the magic elixir that dramatically improved the performance of the 383, you have to look a little closer to determine the facts. In reality, this is a perfect example of engine combinations. Had we made the camshaft swap *before* the intake and exhaust changes, we would have seen a significant improvement in power, but not to the extent that we saw when the cam was changed *after* the intake and exhaust changes. Obviously, the car picked up serious power after the cam swap because it was camshaft limited, but the increase was also in part due to the improved breathing of the intake and exhaust systems additions. While we did not test the engine with the Crane camshaft and the stock intake and exhaust systems, it appears that a cam swap in this engine would be worth more than half of the total gain found with the cam, intake, and exhaust together.

Now we were excited. All we needed were a few subtle horsepower tricks and we could get this Chevelle into the 12s. Since the Chevelle has a very narrow wheelwell, we added a set of 26x8-inch Goodyear slicks on 15x7-inch wheels to the car to prevent tire spin. Up until now, we had been using Mickey Thompson Indy Profile SS tires mounted on the new Cragar Bob Glidden Signature Drag Star 15x7-inch wheels. We also improved the front suspension with a set of Global West Alignment trick aluminum and Delron control arm bushings and added an Air Lift air bag in the right rear to control suspension squat on acceleration.

With all these changes, plus a few more jetting and ignition tune-ups, we decided to travel to Carlsbad Raceway near San Diego, California, to benefit from the better sea-level air to see if the Chevelle could turn some 12s. After two trips to the track and numerous tuning efforts (jetting: 71 primary, 77 secondary; 38 degrees total timing; and a 35 accelerator pump squirter), we were able to pull out a muffled 13.17/103.32-mph pass. While not bad for driving the car in bumper-to-bumper Los Angeles traffic 240-miles round trip to the track with no overheating or loading up, we felt that the Chevelle was still capable of a 12-second pass. As an aside, even when including the dragstrip runs into the fuel mileage, the Chevelle still generates around 12 mpg.

With one more effort at Carlsbad, we removed the front sway bar, screwed the slicks to the wheels, tried colder plugs, cooled the engine down between runs, and concentrated on leaving as hard as possible. Despite our best efforts, and some tuning assistance from our pal Lou Czarnota, the best the Chevelle could generate was only a slightly better mph with 13.17 at 103.55 mph.

Considering that we were still using a

Engine monitoring fell to a slick set of Auto Meter gauges in the form of a Pro Comp tach and Sport Comp gauges to keep track of engine temperature, oil pressure, and amperage.

Neotech's slick electronic tire pressure gauge also came in handy. The gauge display carries pressure out to .2-pound increments.

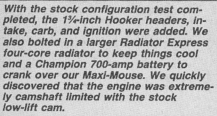

With the stock configuration test completed, the 1¾-inch Hooker headers, intake, carb, and ignition were added. We also bolted in a larger Radiator Express four-core radiator to keep things cool and a Champion 700-amp battery to crank over our Maxi-Mouse. We quickly discovered that the engine was extremely camshaft limited with the stock low-lift cam.

Hooking the car up was no easy task with a high-torque 383. Street-legal Mickey Thompson Indy Profile S/S G60-15 tires made the task much easier, although we eventually had to step up to a set of Goodyear 26x8-inch slicks. We also added Cragar Bob Glidden Signature Drag Star 15x7-inch wheels.

Swapping in the Crane HMV-278 was extremely easy and was accomplished without pulling the engine. We also de-greed-in the cam, finding it perfectly degreed on the first try. The cam, intake, and exhaust combination picked the Chevelle up almost ¾ second, putting it firmly into the high 13s.

MAXI-MOUSE

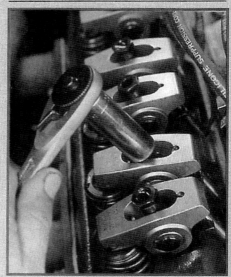

Experimenting with the Crane 1.5 and 1.6:1 aluminum roller rockers after swapping in the ported heads generated slightly better numbers with the 1.6:1 rockers on the intake side and 1.5s on the exhaust. If the expense of the exotic rollers is a little too steep for your budget, Crane now makes a 1.6 stamped steel rocker that is less expensive.

Tuning the car at Jennings Dyno Tune on its chassis dyno also helped immeasurably. Experimenting with accelerator pump squirters and cams, primary and secondary jetting, and total ignition timing all helped the engine's performance significantly.

set of stock 1.94/1.50-inch valve heads on this Maxi-Mouse, a 13.17 e.t. with the Chevelle is respectable. We also noticed that at no time did this engine ever detonate, even with as much as 44 degrees of total timing on nothing more than 91-octane Chevron supreme pump gas. Despite these circumstances, 12-second e.t.'s should be easily attainable with a good set of cylinder heads and maybe a bit more compression—which is what we intend to do with this motor in its next configuration. So stay tuned. In the next few months we'll return with an update on the further adventures of Maxi-Mouse. **HR**

E.T. SLIPS

E.T./MPH	MODIFICATION
15.52/87.28	Baseline w/dead 350 engine, 3.08 gear, Turbo 350, M/T Indy S/S tires
14.90/90.33	New PAW 383, stock cam, iron intake, Q-Jet, cast-iron exhaust, glasspacks, 3.08 gear, B&M trans and converter, M/T tires, shifted at 4500
14.76/90.53	Same as above, w/3.50 gears
14.74/90.66	Same as above, w/Edelbrock intake, Holley 750, Hooker headers, turbo mufflers and complete exhaust. (See story for explanation)
13.79/99.44	Same as above w/Crane camshaft, shifted at 5500 rpm
13.43/102.85	Same as above, w/1.5:1 Crane roller rockers, 38 degrees timing, 71/77 jetting, slicks
13.17/103.55	Same as above, w/colder spark plugs and suspension tuning

PARTS LIST

The majority of the parts listed here are available through Performance Automotive Wholesale (PAW).

Part Number	Component
PAW	383 long-block kit, 9:1 compression, 1.94/1.50 valve heads
113801	Crane HMV-278 hydraulic camshaft (278/290 adv. duration, 222/234 at .050-inch tappet lift, .467/.494-inch lift)
PAW	Lifters, hydraulic
737 PAW	Pushrods, chrome moly
11750-16	Roller rocker arms, Crane, 1.5:1, ⅜ stud
11759-16	Roller rocker arms, Crane, 1.6:1, ⅜ stud
2101	Intake manifold, Edelbrock, dual plane
0-3310-2	Carburetor, Holley 750 vacuum secondary
34-6	Secondary metering block kit, for 3310 Holley
8101	Fuel line, universal, PAW
PK-301	Carburetor jet plate, Redline, primary
PK-302	Calibration kit, for power plate, Redline
M6900	Fuel pump, Carter, mechanical
85800	Air cleaner, Milodon, 14-inch, chrome
E-1650	Air filter, K&N, 14x3 inches
85500	Valve cover, Milodon, tall chrome
3748201	Electronic distributor, Mallory, Unilite
28720	Coil, Pro Master, Mallory
775	Plug wires, Mallory, Pro Wires
2124	Headers, Hooker, 1⅝-inch Super Competition
21105	Turbo mufflers, Hooker, Super Competition
180111	Gears, 3.50:1, Zoom, 12-bolt
PAW	Bearing and installation kit, Zoom, 12-bolt
113001	Transmission, Turbo 350, B&M
20470	Converter, B&M, Torque-Plus 11-inch
80253	Transmission cooler, B&M
80676	Shifter, Quicksilver, B&M
6809	Tachometer, Pro-Comp Memory, Auto Meter
3532	Water temperature, Sport Comp, Auto Meter
3421	Oil pressure, Sport Comp, Auto Meter
3581	Ammeter, Sport Comp, Auto Meter
—	Battery, 650 amp, Champion
—	Gaskets, Fel-Pro, head, intake, exhaust
—	Radiator, four-core custom, Radiator Express
7066	Tires, Indy Profile S/S, G60-15, M/T
—	Slicks, 26x8-inch, Goodyear
4157345	Wheels, Drag Star 15x7, Cragar

The following companies were involved in the Maxi-Mouse test. Most of the parts can be obtained through PAW; if not, the source is listed below.

SOURCES

Cragar Wheels Division of Mr. Gasket Co.
Dept. HR
19007 S. Reyes Ave.
Compton, CA 90221
(213) 639-6211

Global West Alignment Specialties
Dept. HR
5660-B Arrow Hwy
Montclair, CA 91763
(714) 946-7828

Pacific Dunlop-GNB
Dept. HR
1110 Hwy 110
St. Paul, MN 55164
(800) 477-4700, Ext. 52
(Champion battery)

Performance Automotive Wholesale
Dept. HR
21050 Lassen St.
Chatsworth, CA 91311
(818) 998-6000

Radiator Express
Dept. HR
3595 S. Higuera St.
Unit C
San Luis Obispo
CA 93401
(805) 541-6220

Bracket racers are constantly faced with the dilemma of balancing escalating costs against reliability when building engine combinations. It's difficult to build the power needed to put an average car into the 10- to 12-second range and still maintain a budget. For this story, a total of $5000 was determined to be the limit for the entire engine; $2700 was earmarked for the short-block alone. While that might seem excessive to street machiners, it's well within reason for a bracket racer who's looking to crack the 10s.

The engine detailed here will be installed in a '69 Camaro. The buildup starts with Lunati's 406-inch Super Bracket kit, which is based on a production 400 small-block Chevy block-and-crank combination with a .030-overbore to get the extra 6 cubic inches. The 406 provides the torque a bracket racer looks for in the rpm band that is desirable when running an automatic

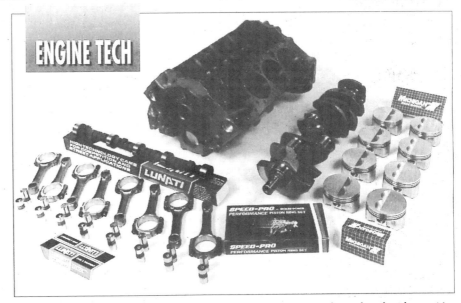

ENGINE TECH

The complete 406-inch Super Bracket package from Lunati can be ordered with or without the block. Options such as Fel-Pro gaskets or high-volume oil pumps can also be included, ensuring that everything is matched for optimum performance.

406-INCH BRACKET SHORT-BLOCK

RELIABILITY AT A REASONABLE PRICE

transmission. Lunati tech adviser Mike Rimmer recommended the addition of a Pro Prepped four-bolt-main, race-ready block and a Pro Series crank option to the basic 406 rotating assembly in order to provide a short-block that would be capable of running longer than a single season, if properly maintained. If the engine were to be built for the street, the basic two-bolt-main block would have been fine.

The Pro Prepped race-ready block cores are each tested to ensure that only high-quality castings are used. After cleaning, each block is Magnafluxed and deburred, and the main caps are shot-peened and inspected. After the deck has been

The Pro Prepped race-ready block comes with the decks milled and is align-honed. Lunati also trues the transmission surface to ensure proper alignment.

Each cylinder is torque-plate-honed and retains enough crosshatch to properly seat the rings. After all of the machine work is completed, the block is scrubbed clean. Note the water-transfer passages just above and between each cylinder. The plugs add additional strength to the deck surface yet provide adequate water flow.

trued and the transmission mounting surface has been squared, the block is torque-plate-bored and honed.

Before the block visits the boring bar, plugs are installed in the steam holes and oil-galley holes. The lifter galley is deburred and painted to ensure cleanliness and free oil flow. The block is clearance-ground if longer rods or a stroker crank is being used, and bronze freeze plugs are installed after the block is painted. Since the block used for this story was bored only .030 inch, Rimmer felt that strengthening with block filler wasn't necessary, but he did recommend the Pro Series cast crank, primarily because of the heat-treating process involved. Additionally, the Pro Series crank is held to closer tolerances, which is necessary in all-out racing applications.

A Pro Series crank is stroke-corrected so that every throw is *exactly* the same. Then it is shot-peened, Magnafluxed, detailed and trued. After being precision-ground and flange-trued, the oil holes are chamfered and the crank is micropolished. The Pro Series crank adds $275 to the kit price.

A Custom Grind 400 crank comes standard with the 406-inch assembly, and Lunati also offers a 4340 forged steel crank, which carries a one-year guarantee against breakage when used in unblown applications. Rimmer pointed out that the 4340 crank is more durable and stronger than the standard crank, but since it costs over $1000, the budget for this engine decided against it.

High-compression-domed or flattop pistons are available from Lunati, but since the 406 is equipped with 350 Chevy 5.7-inch Stage 3 Lunati connecting rods—which are .135 inch longer than the 400 rods—a flattop piston (Lunati part No. 92102-3) with a relocated piston pin and 1/16-inch rings was selected. Each piston weighs 491 grams and allows nearly 12.5:1 compression with a .010-inch deck clear-

Block surfaces that would interfere with proper rod and crank rotation must be ground away when longer rods or stroker cranks are used. Sharp edges are also removed to prevent stress risers.

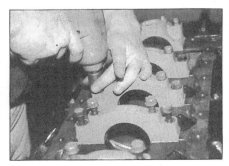

An impact wrench is fine for removing main-bearing caps, but use only a high-quality torque wrench when assembly begins. After checking all oil clearances, sequence-tighten each nut and bolt.

When the lifter galley is being painted, Lunati uses inverted umbrella seals that block paint from running into the lifter bores to prevent problems with lifter clearance. Also, note that masking tape has been placed over the new cam bearings to prevent paint from ruining them.

Installing new freeze plugs is good insurance against problems that could occur later. Bronze plugs are used in race-ready applications, but O.E.M.-type steel replacements are fine for most street applications.

GREAT READING ON GTO's & FIREBIRDS FROM:

CarTech
1 800 551 4754

Clean the block with soap and water and then spray it with a good water-repelling agent such as WD-40. Wipe away any surplus with a clean cloth and use high-pressure air to remove any lint or other small particles. You can't overdo cleanliness.

Lunati custom-grinds each Pro Series crank. Stroke correction and indexing are also part of the procedure.

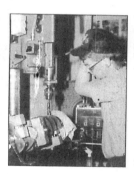

Lunati internally balances the 406 so that neutral-balance flexplates and harmonic balancers from non-400 small-blocks can be used.

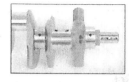

After each oil hole is chamfered, the crank receives a micropolishing, leaving a mirror-like finish on the journals. When the Pro Series crank has been heat-treated, the counterweights will have a dull, reddish, hazy appearance. Heat-treating creates additional surface hardening to improve crank durability.

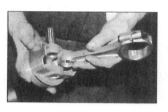

Lunati's Stage 3 reconditioned rods feature ARP rod bolts, polished beams and resized big ends. The addition of Ampco 18 bronze bushings allows the rods to be used as full-floaters. This requires that spiral locks be used in the pistons to retain the lightweight Lunati pins. Oil holes are added to the small ends of the rods for proper oiling.

The pin is placed higher in the flattop piston for the 406 assembly, allowing for 5.7-inch 350 rods and increased power. Note the 1/16- and 3/16-inch ring lands. Narrow-width rings provide adequate seal and maintain proper oil control yet create less friction.

Double-check deck height, piston-skirt clearance, block-to-rod clearance and cam-to-rod clearance. The 406 with flattop pistons needs approximately .010-inch deck clearance, .006- to .008-inch skirt clearance and about .060-inch rod-to-block and rod-to-cam clearance.

Side clearances of approximately .020 inch between the rods and crank flanges are also critical in race engines. Improper side clearance restricts oil flow and generates heat, but excessive clearances can also rob horsepower and lead to damaged parts.

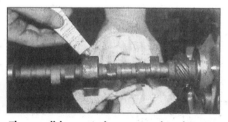

The small-base-circle cam used in the 406 ensures that the piston and rod assemblies stay clear of the bottom of the cam lobes. Most camshaft failures are the direct result of improper lubrication upon initial start up. Lubricate both the cam and the lifters and prelube the engine before firing it up for the first time.

ance and 64cc combustion chambers.

As with the other components, the Stage 3 rods are Magnafluxed and shot-peened, and they're checked for bends and twists before the beams are ground and polished. The beams are then shot-peened again to provide maximum strength, and Lunati's rod shop hones the big end within plus or minus .0001-inch tolerance. The shop then installs Ampco 18 bronze bushings and ARP heavy-duty rod bolts.

Oil holes are drilled into the small end of each rod, and the rods are then grouped in weight-matched sets to ensure precision balancing. Lunati also offers 4340 forged-alloy Pro Mod connecting rods for $297 more, but the budget again determined that the Stage 3 rods be used. Lunati says that the Stage 3s are good for up to 7000 rpm, which is more than this bracket motor should ever see.

As with almost any engine, cam selection is where all the effort can pay off. Rimmer recommended a Lunati solid-lifter cam (part No. 40136) for this application. It's a small-base-circle grind and features 249-degree intake duration and 259-degree exhaust duration at .050-inch tappet lift. Lift at the valve is .543 inch on the intake and .561 inch on the exhaust using 1.5:1 rocker arms. Lobe separation is 106 degrees, and the advertised duration is 293/303 degrees. Lash is set at .018/.020. Since the Camaro for this bracket effort weighs 3250 pounds, the tighter lobe separation and a 3000-stall converter should get the car off the line without transferring too much shock to the drivetrain. With good cylinder heads and the appropriate exhaust system, this short-block has the capability to make 450 horsepower.

On the budget front, the Lunati kit—even with options—met the goal. The basic Super Bracket 406 assembly was priced at $1715. Upgrade options to the Super Bracket kit included the Pro Prepped block for $695 and the Pro Series crank for $275, bringing the total to $2685. Lunati also carries a full range of cylinder heads, high-volume oil pumps and pickups, oil pans, complete Fel-Pro gasket kits and all of the other small parts that will be needed to complete this engine. **HR**

SOURCE

Lunati Cams
Dept. HR05
P.O. Box 18021
Memphis, TN 38181-0021
901/365-0950

SCRATCH-BUILT 377 part 1

DO YOU KNOW WHAT IT TAKES TO BUILD YOUR FIRST ENGINE, STARTING WITH NOTHING?

By Mike Magda

"Everyone's building 383s," said a longtime reader. "Why don't you build a really hot 377 and be different?"

"Okay, let's do it," we replied. "What do we need?"

"That's easy—a 350 crank and a 400 block."

"And what else?"

"What do you mean?"

"What else do we need to build an engine?"

"The usual stuff: carburetor, intake manifold, cylinder heads...."

"Which heads?"

"I don't know. How about Edelbrock?"

"Fine. Which Edelbrock heads?"

"Uh, the best ones."

"How about valves and springs?"

"Don't they come with 'em?"

"Some do, some don't"

"Jeez, why are you guys making this so difficult? Just build a regular Chevy small-block."

Building an engine requires more than an idea. We're not talking just parts, of which there are many that need to be gathered. We're also talking about research and preparation to make the right decisions so that all the parts work together.

With the challenge posed to us to build a 377, we decided to start from scratch and let you follow along as

SHOPPING SUGGESTIONS

Research and a lot of patience will be your biggest assets when building an engine from scratch. You have the power to make all the right decisions, compared to someone else who is forced to work with existing parts.

We started by reviewing dozens of catalogs and paying close attention to parts compatibility. Just because the parts description says "Chevy small-block" doesn't necessarily mean it will fit. The first concern is mating the crankshaft to the block. Chevy uses both an old-style two-piece rear main seal and upgraded one-piece seal. It is possible to install an early two-piece-style crank in a late-model block with a rear-seal adapter. We'll discuss other compatibility problems throughout this article.

Take a good look at the engine kits various aftermarket mail-order firms offer. These are designed with parts chosen to work with each other. Discuss your application with the tech reps to get the right compression ratio, cam profile, and so on. Engine kits come in numerous configurations, starting with basic gasket, ring, and bearing sets. More extensive offerings will include pistons, the crankshaft, camshaft and lifters, pushrods, oil pump, and a timing set. Many options are available to suit all pocketbooks, including forged pistons, moly rings, and double-roller timing chain. Study the advertisements in *Chevy High Performance* to get a good idea of the types of kits available.

Most kits will not include everything. There are numerous details that have to be addressed after all the major parts are selected.

Catalogs are a great source of information and will help you plan an engine on paper. Most have technical advice and excellent parts descriptions to help with compatibility questions.

Engine kits come in various configurations. This kit has most of the major internal parts, but you would still have to gather a block, induction system, ignition, and numerous small finishing parts.

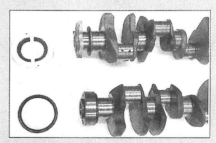

Compatibility is the major concern when building an engine from scratch. Not all small-block crankshafts fit, as Chevrolet changed from a two-piece rear seal (top) to a one-piece.

SCRATCH-BUILT 377, PART 1

we develop a plan and continue through the parts selection, machine work, assembly, and dyno testing.

Under ideal circumstances, the cubic-inch displacement would be determined after evaluating the purpose of the engine as well as other factors such as cost, availability of parts, rules restrictions for race applications, and so on. A 377ci displacement has already been established, so we will make the most of that decision.

The traditional approach to achieving 377 cubic inches is installing a 3.48-inch-stroke crankshaft from a 350 block in a 400 block that has been bored 0.030 over to 4.155 inches. A stock bore for a 350 is 4.000 inches, so a 377 is a 350 with .155-inch-larger bores. The catch is that a 350 crank just doesn't bolt in a 400 block. A 350 crank has 2.45-inch main journals, while a 400 block is designed for 2.65-inch main bearings. Aftermarket bearing spacers are needed to fit the 350 bearings in a 400 block.

A 377 is more rev-happy and suited to high horsepower than a 383, which is achieved by taking a long 3.75-inch-stroke crankshaft from a 400 engine, regrinding the journals, and dropping it in a .030-over 350 block. A 383 will pump out more torque for the extra cubic inches but can't be twisted as high as a 377. A 377, therefore, is more suited to high-end bracket racing than performance street use. A 377 will work best when a lot of horsepower is built into the high-rpm ranges, then setting up the car with very low gears.

Since we would have to approach 7,500 rpm to achieve maximum potential, that kind of stress narrows the list of parts that will live in that environment. To build a 377 right, you have to go all-out.

That doesn't mean you can't build a street 377. A big-bore 350 will perform better than a stock 350 simply because the valves will be unshrouded a little more from the cylinder walls. With better breathing, you

BLOCK

The key to a good cylinder block is the machine work and preparation. Most two-bolt blocks will be good for up to 350 hp. Anytime you want more than 400 hp, a four-bolt block should be used, especially if you intend to race.

We chose a CNC Bow Tie block from GM Performance. This is a siamesed-bore case, meaning that the cooling passages do not flow completely around the cylinders. The four-bolt steel caps have 20-degree splayed outer bolts, and the block comes with main studs. The block also has plenty of racer options, including dry-sump oiling access, enlarged lifter bosses, and priority main oiling.

The CNC Bow Tie block comes with a 9.250-inch deck height, but our pistons require an 8.99-inch height for the proper compression ratio. We'll follow all the block prepping in an upcoming issue. We mounted our block on a TD Performance engine stand that is equipped with a JAZ Products drip tray. When needed, the engine will be covered with a Moroso engine bag. We'll be running Red Line oil during the main dyno tests.

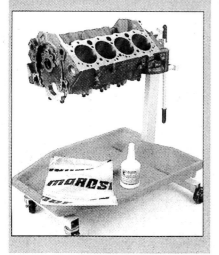

BEARINGS

Engine builders do have options when selecting bearings. There are different designs in the levels of eccentricity and crush, in addition to bearings that are made from different materials. Certain bearings are best suited for specific applications. Study the bearing catalogs closely. Don't try to order a racing bearing for a street car. If you have a racing or special application, then seek advice from a tech rep.

We chose the V-series line of Clevite 77 bearings. This is basically the old Vandervell parts line. V-series bearings have low-to-medium eccentricity and a hardened steel back. We also ordered our main and rod bearings with an "X" suffix, meaning we get an extra .001 oil clearance for a high-revving motor.

A generous supply of assembly lubricant is also helpful. Some of our favorites come from Red Line, Permatex, and ARP.

BLOCK ACCESSORIES

Cylinder blocks rarely come from the machine shop complete. Make sure you have soft plugs (sometimes called freeze plugs), head dowels, bellhousing dowels, pipe plugs for the oil galleries (if you don't want to use soft plugs), and an oil filter adapter (two bolts needed). Bellhousing dowels can be ordered extra-long for more secure mounting of the scattershield or transmission. Our block accessories came from Pioneer, Mr. Gasket, GM Performance, and

Moroso. We also ordered an extra-long Fram oil filter for more protection during dyno tests. This filter may interfere with the chassis or headers in some cars.

Other options for a racing block include deck plugs. These ¾-inch NPT plugs are installed in the water passage holes to stiffen up the deck and reduce cylinder wall distortion. The holes must tapped, and block machining is necessary after installation.

Another trick is an oil return screen kit made up of stainless-steel mesh screens that are fitted over the holes in the lifter galley to trap metal fragments. We also plan to install oil restrictors that reduce the flow of oil to the rocker arm assemblies, leaving more for the mains and rods. (Note: oil restrictors cannot be used with hydraulic lifters.)

Plenty of block work is ahead for us. We'll smooth out the lifter galley and all the rough edges with Eastwood carbide burrs and grit rolls from Standard Abrasives before applying VHT paint.

Finally, don't forget the dipstick. We picked up a chromed model from TD Performance.

SCRATCH-BUILT 377, PART 1

might be able to tone down the cam timing slightly to lower the torque curve and pull better on the low end without sacrificing top-end power. Again, if you gear a 377 properly, it will work on the street. But part of our challenge was to build an engine from scratch. If we were starting with a clean sheet of paper to build a street small-block engine, we would go with a 383 or 406 to pull as much torque as possible for better driveability and more neck-snapping throttle response. However, since our mission is to build a 377 from scratch, we will build a high-strung twister for bracket or open-road racing.

The best advice we can give when starting an engine project is to get good advice from the engine builder or machine shop and the tech representatives of the aftermarket performance companies. Don't just talk to them once. As you develop a plan, you'll discover that their recommendations will change as the direction of the project narrows. Tech reps like to have everything else in place before selecting the right part from their catalog, so there's a lot of give-and-take along the way. For example, the camshaft rep will want to know the carb size you're using, and the carburetor rep will want to know which cam you've got before making any recommendations. The piston rep will need to know which head will be used and the desired compression ratio, as well which cam you order. The cam rep will then want to know how much the head flows.

As you can see, it can get to be quite a juggling act, but the extra legwork and big phone bill will pay off in the long run if you develop a good relationship with the tech reps. A word of advice: Don't waste their time with idle chat or vague questions. Be direct and have a plan. Know everything there is to know about your car and its purpose.

MORE STUFF FOR THE BLOCK

You'll need a starter. We chose a high-torque mini starter from GM

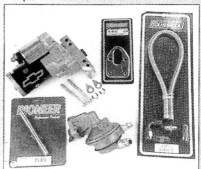

Performance. It will be secured with Stage 8 locking safety bolts. If you're running an electric fuel pump, you'll need a block-off plate, gaskets, and bolts to cover the opening in the block. If you're running a mechanical fuel pump, you'll need a mounting plate, gaskets, bolts, and a fuel pump pushrod, plus fittings and fuel line to the carburetor. Fuel for our dyno tests will be supplied by an electric pump directly to the carbs, but future plans may call for in-car testing. We would use a Carter racing pump with fittings already brazed in place and Russell plumbing.

HEAD BOLTS

Aftermarket heads often require special head bolts, so check with the manufacturer. The Edelbrock heads used standard-sized head bolts; however, ARP manufactures a variety of head bolt styles. We chose the hex-head stainless steel model. The Bow Tie blocks have blind bolt holes, so it's always a good idea to check the lengths to prevent the bolts from bottoming out. Also, use a good thread lubricant to ensure correct torque readings.

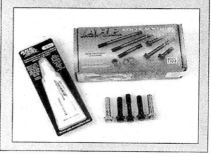

ROTATING ASSEMBLY

A high-revving engine needs a steel crankshaft. Cast iron will not hold up under the stress of 7,500 rpm, and weight will impede performance dramatically. We chose a 1053 steel crank from GM Performance. This crank has the old-style two-piece rear seal design and is nitride-treated to increase journal hardness. The GM crank was recently improved, and we're eager to test one of the first units.

Crankshafts don't always come complete. You need a pilot bearing (we chose a roller model from Pioneer) and a woodruff key (again, we went with Pioneer). Also, the rear main seals should come in a gasket kit.

The crankshaft should be balanced with the flexplate (or flywheel) and harmonic balancer. We chose a B&M flexplate (mounted with ARP bolts) and TCI's new Rattler balancer (secured with an ARP bolt).

RODS & PISTONS

Now the variables start coming into play. Choosing the right rod/piston combination takes time. The rod must meet the demands of the engine. Luckily, many new rod manufacturers have come on the market with excellent forgings that are secured with quality fasteners. We chose the new Diatron line from Klein Engines. Made from 4340 steel, these I-beam rods feature

ARP bolts rated at 230,000 psi. We also went with 6-inch-long rods for reasons explained elsewhere in this magazine.

Since we knew the stroke and rod length, we could order the piston in addition to the cylinder head. We wanted a compression ratio around 12.5:1, so the experts at Wiseco pistons suggested its drag race line with a 1.250 compression height and 12cc dome. The block will have to be cut down to an 8.990 deck, and we may have to mill the 70cc chamber heads to get the desired compression ratio. The pistons come with 4140 chrome-moly wristpins and spiro lox. We wrapped the pistons with Speed-Pro plasma-moly rings.

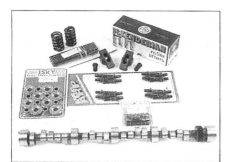

CAMSHAFT

Once you've established a purpose for the engine, displacement, cylinder heads, and compression ratio, the camshaft can be ordered. Perhaps the biggest mistake made in building any engine is installing an overly radical camshaft.

We're going to have big-horsepower, upper-rpm numbers, so big is just right for our 377. Shaver and Richard Iskenderian chose a custom roller-rocker grind featuring 264/272 duration at .050 with 108 lobe centers. Shaver wants to run Isky's 1.65:1 roller rockers on the intake and 1.5:1 rockers on the exhaust, resulting in .714 lift on the intake and .645 on the exhaust.

Mixing valvetrain components is not a good idea, so we went with all Isky products, including .100-inch longer-than-stock pushrods, springs (230 pounds at 1.970-inch height), titanium retainers, and valve locks. We tossed in ARP's 7/16-inch rocker studs.

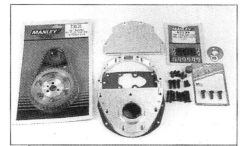

CAM TIMING

Dyno testing a high-strung engine means playing with all the variables. We want the option of quickly changing camshaft advance or retard, so we chose a Manley timing set and three-piece timing cover. The Manley timing set features a race roller chain and fully adjustable two-piece camshaft sprocket. The cam can be adjusted by 10 degrees on either side. The kit also comes with a Torrington bearing to prevent block wear. The Manley timing cover features a removable top half to allow for quick adjustment without removing the entire cam cover. The kit comes with all the necessary hardware.

OILING

Moroso offers a complete line of oiling systems with all the options possible. Oil is both friend and foe

to an engine. A good layer of lubrication is needed for friction-free running and cooling, but too much oil can pull horsepower away from the engine. For our 377, Moroso suggested its new Stage II Pro Eliminator 6-quart oil pan that features a full-length, unidirectional windage tray screen. A block-high scraper is included. The pan also features a big kickout on the passenger side to reduce windage and extra-thick pan rails for improved sealing. It comes with a magnetic drain plug, and a special Moroso mounting kit is required.

Moroso also suggested its high-performance oil pump. Additional parts required include an oil pump mounting stud, oil pickup, and oil pump driveshaft.

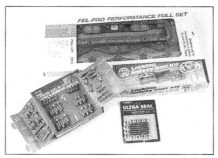

SEALING & OTHER HARDWARE

Even though you get gaskets and mounting hardware with some individual items, it's a good idea to purchase complete gasket and fastener kits. You can always use the leftovers in future projects or during routine teardowns. We picked up a Fel-Pro gasket kit to seal the engine. For fasteners, we had the option of picking between an ARP dress-up kit and a Stage 8 safety kit. We also had a set of Mr. Gasket Ultra-Seal header studs for when the engine goes to the dyno room.

CYLINDER HEADS

The Edelbrock Victor Jr. heads come with 2.08/1.60 stainless-steel valves. These heads are designed for out-of-the-box racing applications and have fully CNC-machined raised exhaust ports and combustion chambers. The intake side is CNC machined about 2 inches inside and matched to a

1205 Fel-Pro gasket. Intake runner volume is 212cc, while the combustion chamber is a free-breathing 70cc. The heads need pushrod guideplates and valve seals in addition to the Isky equipment we've already gathered for complete assembly.

CARBURETION

Carburetors are becoming more complex every day. Carb tuners have developed numerous performance modifications that can be tuned to every kind of application. We called on Barry Grant for his advice, and he sent out a pair of carbs for testing. The recommended 750-cfm Stage 3 drag carb will probably work best throughout the power range, but for peak top-end performance he asked us to try a Dominator, since the cam looked pretty healthy. We have a couple other Barry Grant carbs on the bench that may see action as well.

The RT Stage 3 carbs are designed for fast throttle response and feature custom fuel circuits, removed choke horns, custom-finished venturis, four-corner idle, radiused air entry, machined metering surfaces, and dual-feed double-pumper bowls.

Russell Performance will help plumb the carbs with its competition inlet lines, fuel pressure gauge, fuel filter, and fittings.

SCRATCH-BUILT 377, PART 1

SCRATCH-BUILT 377 PARTS CHECKLIST

DESCRIPTION	MANUFACTURER	PART NO.

BLOCK

CNC cylinder block	GM Performance	24502503
Main bearing caps	GM Performance	with block
Main studs & nuts	GM Performance	with block
Camshaft bearings	Clevite 77	SH1349S
Main bearings	Clevite 77	MS909VX
Soft plugs	Pioneer	83001
Debris screen	Moroso	25000
Head dowels	Pioneer	839004
Bellhousing dowels	Moroso	37932
Deck plugs	Moroso	37800
Oil restrictors	Moroso	22016
Paint	VHT	n/a
Oil filter mount	GM Performance	3952301
Oil filter mount bolts	Shaver Racing	n/a
Oil filter	Fram	PH373
Dipstick	TD Performance	4957
Pipe plugs	Mr. Gasket	3806 & 3807

BLOCK ACCESSORIES

Starter	GM Performance	12363128
Starter bolts	Stage 8	8996
Starter shield	TD Performance	9796
Fuel pump	Carter	M60969
Fuel pump pushrod	Pioneer	839036
Fuel pump mounting plate	TD Performance	2310
Gaskets	Fel-Pro	in kit
Mounting plate bolts	ARP	in kit
Fuel pump fittings & hose	Russell	4103

CRANKSHAFT

Steel crankshaft	GM Performance	3941184
Woodruff key	Pioneer	839009
Pilot bearing	Pioneer	873008
Rear main seal	Fel-Pro	in kit
Balancer	TCI	870001
Balancer bolt	ARP	234-2501
Flexplate	B&M	20230
Flexplate bolts	ARP	200-2902

ROD/PISTON

Pistons	Wiseco	K083A3
Piston pins	Wiseco	w/pistons
Pin locks	Wiseco	w/pistons
Rings	Speed-Pro	R9346 3
Connecting rods	Klein	n/a
Rod bearings	Clevite 77	CB 663VX

CAMSHAFT & VALVETRAIN

Camshaft	Isky	201-RR8620
Roller lifters	Isky	1241-LSH
Pushrods	Isky	1235-L
Roller rockers (1.65:1)	Isky	204-65-716
Roller rockers (1.5:1)	Isky	204-716
Polylocks	Isky	w/rockers
Rocker studs	ARP	200-7201
Valvesprings	Isky	9375-85
Titanium retainers	Isky	90-TI
Valve locks	Isky	VL-200+50
Timing cover	Manley	42129
Timing cover bolts	Manley	w/cover
Cam button	Manley	w/cover
Timing cover gasket	Fel-Pro	in kit
Timing gears & chain	Manley	73131
Camshaft retainer bolts	ARP	234-1001

CYLINDER HEAD

Cylinder head	Edelbrock	7700
Intake valves	Edelbrock	w/heads
Exhaust valves	Edelbrock	w/heads
Valve seals	Shaver Racing	n/a
Pushrod guideplates	Edelbrock	9661
Head gaskets	Fel-Pro	in kit
Head bolts	ARP	434-3601
Valve covers	GM Performance	10185064
Valve cover studs	ARP	400-7603
Breathers	K&N	62-1150

OILING SYSTEM

Oil pan	Moroso	21237
Oil pump	Moroso	22100
Oil pump stud	Moroso	38150
Oil pump pickup	Moroso	24175
Pan studs	Moroso	38385
Oil pump shaft	Moroso	22070

COOLING

Water pump	Moroso	63539
Pump bolts	Moroso	w/pump
Pump gaskets	Moroso	w/pump
Thermostat housing	TD Performance	6003
Housing bolts	ARP	in kit
Water restrictor	Mr. Gasket	6126

INDUCTION

Carburetor	Barry Grant	RT-4779-S3
Fuel inlet assembly	Russell	4111
Fuel filter	Russell	4530
Fuel pressure gauge	Russell	5035
Air filter	K&N	61-4010
Air filter stud	Mr. Gasket	w/filter
Intake manifold	Edelbrock	2900
Carb studs	ARP	400-2401
Pipe plugs	Mr. Gasket	6352
Adapter (for carb test)	Barry Grant	110012
Gasket	Fel-Pro	1205
Throttle return spring	TD Performance	2385

IGNITION

Distributor	Mallory	8448205
Ignition control	Mallory	692
Coil	Mallory	29625
Spark plugs	Champion	C59YC
Spark plug wires	Mallory	937
Terminal kit	Mallory	669
Distributor hold-down	Mr. Gasket	6197
Hold-down bolt	Stage 8	8903

MISCELLANEOUS

Gasket kit	Fel-Pro	2802
Engine accessory bolt kit	ARP	534-9601
Thread sealer	ARP	100-9904
Engine stand	TD Performance	9958
Engine stand bolts	Shaver Racing	n/a
Engine stand tray	JAZ Products	720-000-06
Assembly lubricant	Red Line, Permatex	80302 & 81950
Oil	Red Line	20-50
Torque wrench	Craftsman	44545
Cleaning brushes	Mr. Gasket	5192
Oil primer	TD Performance	9534
Storage bag	Moroso	99400
Cam checker tool	Pioneer	46417
Degree wheel	Pioneer	890103
TDO plate	Pioneer	488060
Ring compressor	Powerhouse	104155
Ring expander pliers	Powerhouse	105060
Deck bridge	Powerhouse	101310
Crank gear installer	Powerhouse	101525
Valve spring compressor	Eastwood	2500
Grit rolls	Standard Abrasives	various
Carbide burrs	Eastwood	3154
Precision tools	Central Tools	various
Balancer installer	TCI	924000
Soft plug installer	Manley	41726
Engine safety bolt kit	Stage 8	9011

TOOLS
If you're going to build an engine from scratch, you'll need plenty of tools, including some specialty items. Besides a set of standard handtools, including ratchets, wrenches, screwdrivers, pliers, and sockets, you'll need a torque wrench. Our trusty standby is a Craftsman Digitork. You just dial in the required torque setting and tighten until it clicks.

Degreeing in the cam is another important step. We like the big Pioneer degree wheel and cam-checker tool for easy operation. A Pioneer TDC plate helps find top dead center. Another necessary tool is a ring compressor. Different models are available, but the best kind is a tapered compressor that gently squeezes the rings as the piston is installed. The downside is that they only work for a specific bore size.

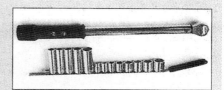

Other tools we'll be using include a Moroso soft plug installer, Powerhouse crank gear installer, a TCI balancer installer, Eastwood valvespring compressor, Powerhouse ring installer, and TD Performance oil primer. A good set of cleaning brushes came from Mr. Gasket. We'll use precision measuring instruments from Central Tools.

Even though this is a dyno project, we developed a model car that this 377 will power. We came up with a 3,000-pound Camaro that will run 10.5-inch slicks. The car will have a TH400 tranny, and gearing will be determined following the dyno tests. Nitrous will not be used.

Finally, we wanted to use off-the-shelf parts. This would present quite a challenge, since the engine would be spun so fast, but with careful research we came up with a strong combination.

With those goals in mind, we called Ron Shaver at Shaver Racing to build the engine and guide us through the parts selection. In our first meeting he suggested using the new Edelbrock Victor Jr. head. He helped develop this CNC-machined race head and felt it would be a strong choice. Shaver didn't like the idea of using bearing spacers in a 400 block, so he suggested a strong four-bolt Bow Tie block that could be bored to 4.155 yet still had the small crankshaft main bearings. He noted that GM Performance has a CNC version that would reduce his time in the machine shop as well.

With the heads and block ordered, we were on our way. What follows is a rundown of the other parts we chose, along with tips and advice we received. Pay close attention to the small items, because you don't want to be running down to the speed shop at the last minute. This 377 will be a racing engine, so some parts are not applicable to the street. Also, we still have to gather exhaust components and accessories such as pulleys, an alternator, and accessory brackets. In the next story we'll follow the machine work and basic blueprinting. In following stories we'll have the step-by-step assembly and dyno testing. Stay with us. ■

ACCESSORIES
The neatest-looking valve cover around is the NASCAR-influenced cast-aluminum model from GM Performance. We picked up a set and hope to have them finished in the traditional black with red trim paint. K&N offered two styles of breathers: One will require a grommet but can be removed often for adding oil. We'll use ARP studs to mount the rocker covers.

Although not necessary on the dyno, should this engine be set up for the street we would need an air cleaner. A good choice is the Flow Control from K&N that comes with a 4-inch-tall element and carb stud.

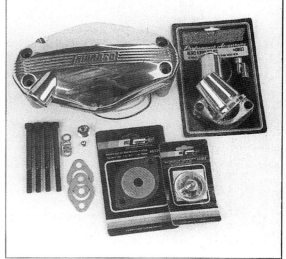

COOLING
An electric water pump is preferred for dyno testing for easy removal. It's also recommended for high-end bracket racers. We chose Moroso's spiffy chromed model that pumps 19 gallons a minute. This lightweight pump comes with all the necessary mounting hardware.

We went with a polished water neck from TD Performance. Mr. Gasket supplied both a thermostat and water restrictor for testing.

INTAKE MANIFOLD
Edelbrock has a Victor Jr. intake manifold already hand-blended to match the heads' intake ports. This single-plane design is good for 8,000 rpm. The manifold has a couple of holes that will be filled with Mr. Gasket pipe plugs. We also have an adapter for the Dominator carb from Barry Grant and a throttle return spring from TD Performance. Gaskets will come from Fel-Pro, and carb studs and mounting bolts will come from ARP.

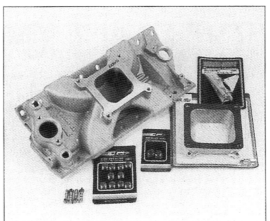

WE FOLLOW THE MACHINE WORK ON A HIGH-REVVING SMALL-BLOCK AT SHAVER RACING

By Mike Magda

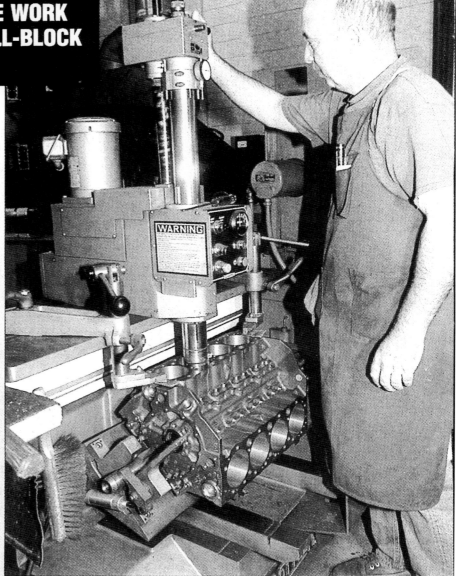

In Part I, we introduced you to a 377ci engine project at Shaver Racing. The 377 is based on a Bow Tie block bored out to 4.155 and fitted with a 3.48-inch GM steel crankshaft. A 377 engine is meant to be spun high, so we gathered some of the best parts around to make sure it lives and performs up to racing standards.

The first step in any engine project is block preparation. The science of block preparation has expanded dramatically in the past few years. Race teams are looking for any advantage, so upwards of 100 to 200 man-hours of work on the block are becoming routine in some race shops. The block is not a source to add more horsepower, but the right work can free up more horsepower by reducing frictional losses. Proper prepping will also increase durability.

The advent of CNC machining on cylinder blocks has cut down time in the machine shop. Some of the duties no longer required for all-out racing blocks include installing splayed four-bolt main caps, blueprinting machine surfaces such as the oil-pan rails, and checking the head dowel locations.

This project features the new racing Bow Tie block from GM Performance Parts. Listed as PN 24502503, this block is CNC-machined from raw casting No. 10051184. All traditional racing modifications are provided by the factory, such as drilling and tapping oil galleries for dry-sump systems. The Bow Tie block is a siamese-bore design, meaning there are no water passages between the cylinders, thus allowing bigger bores. The four-bolt block comes with steel main caps. The intermediate caps have 20-degree splayed outer bolts, while the inner studs are spread out .210 to allow the main caps to be align-bored out to accept a 2.65-inch-journal crankshaft (same as the 400 crank). Other tricks on this block include enlarged cam bosses front and rear to allow for bigger bearings, enlarged lifter bosses, an early-style two-piece rear main seal, and priority main oiling.

PHOTOGRAPHY: MIKE MAGDA

This block is available with either a standard 9.025 deck height or a tall-deck version at 9.150.

Ray Bates, a machinist with 35 years of experience, handled the work on the Bow Tie block at Shaver Racing. He prefers to clean, Magnaflux, and pressure-test all blocks—even new ones—before starting any machine work. Bates also sonic-tests the cylinder walls of blocks; however, CNC Bow Tie blocks come with that information. Bates then deburrs the block and performs any necessary mill work, such as shortening the lifter bosses and machining the front cam boss to clear timing gears and chains.

SCRATCH-BUILT 377

PART II

Bates installs his own oil restrictors to impede the flow of oil to the top end, concentrating the oil on the mains. He then pins the soft plugs after installation.

The block is bored out to 4.150 inches, and then the deck is cut down to 8.99 to work with the Wiseco pistons. Bates then surfaces the bottom of the main caps to ensure a flat fit before align-honing the mains. The final chore is to power-hone the cylinders out to the desired bore. As we near completion, the finishing touches include a wash and installation of the cam bearings.

Follow along as we watch Bates handle all the machine work on this high-revving small-block. In upcoming segments, we'll assemble the Edelbrock cylinder heads, assemble the engine, and dyno-test it.

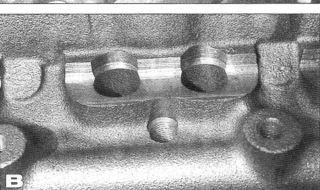

Here are the lifter bosses in a stock CNC Bow Tie block (A). They must be milled down (B) on a Bridgeport machine to clear the links on the Isky roller lifters that we'll be using. This step is not necessary for hydraulic or solid lifters. Note that the oil drainback holes are threaded for easy installation of pressure relief tubes.

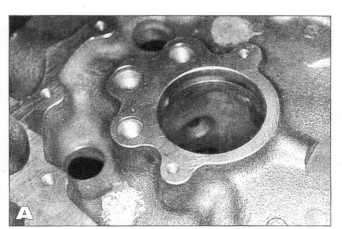

Here's the stock cam bearing boss on the front of a CNC Bow Tie block (A). It's very thick in order to support larger bearings, if desired. However, sometimes the extra metal can interfere with the timing chain or gears. Again using a Bridgeport, Bates mills down the top half of the boss .200 inch (B) for extra clearance.

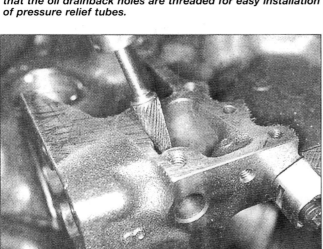

Deburring the block is a time-consuming chore but is well worth the effort to save an engine. Deburring removes sharp casting flashes that could break off and find their way into the oil system. The work is easily done with a die grinder fitted with Eastwood carbide burrs.

Bates drills a small hole through a ¼-inch pipe to restrict the oil in the rear oil galleries.

SCRATCH-BUILT 377, PART II

The pipe plug is installed...

...in front of a ⅜-inch pipe plug. This trick is used only on the side galleries, not the center one.

The Pioneer soft plugs are installed and held in place with pins.

Moroso deck plugs are installed to add rigidity to the block. The water holes have to be tapped with a ¾-inch pipe tap before installation.

With the pipe plugs in place, the deck is milled .026 on a Sunnen surfacer. An 8.99-inch deck height is required to work with the Klein 6-inch rods and the Wiseco pistons, which have a 1.25-inch compression height.

The Moroso oil pump is torqued down on the rear main cap during line honing to simulate the pressure the cap endures during operation. Note the intermediate steel main caps that have 20-degree splayed outer bolts.

The block is set up on a Rottler boring bar and bored out to 4.150 inches.

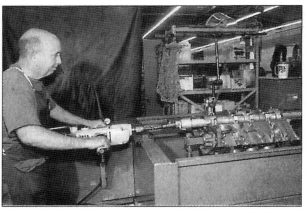

Bates cleans up the main bores with a quick align-hone.

The intermediate caps require two kinds of washers. The flat washer on the left is used with the studs, while the countersunk washer on the right is used with the outside bolts.

The bores are checked with a dial bore gauge.

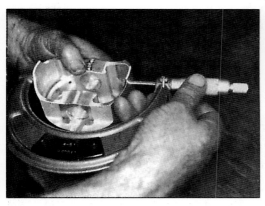

The Wiseco pistons are measured before setting the power-hone machine.

With the block almost complete, Bates turned his attention to balancing the GM Performance crankshaft. The block will be washed and fitted with Clevite 77 bearings before being shipped to the engine assembly room at Shaver Racing. ■

The block is fitted with torque plates during the power-hone to simulate the stresses of the heads being torqued down. This ensures a perfectly round cylinder. Bates uses a rough stone for most of the work before switching to a 625-grit for the finishing work.

SOURCES

Eastwood Company
Dept. CHP
580 Lancaster Ave.
Malvern, PA 19355
610/640-1450

GM Performance
Dept. CHP
See your local GM dealer
Call 800/577-6888

Moroso
Dept. CHP
80 Carter Dr.
Guilford, CT 06437
203/453-6571

Pioneer Automotive
Dept. CHP
5184 Pioneer Rd.
Meridian, MS 39301
601/483-5211

Shaver Specialty
Dept. CHP
20608 Earl St.
Torrance, CA 90503
310/370-6941

Wiseco Pistons
Dept. CHP
7201 Industrial Park Blvd.
Mentor, OH 44060
800/321-1364

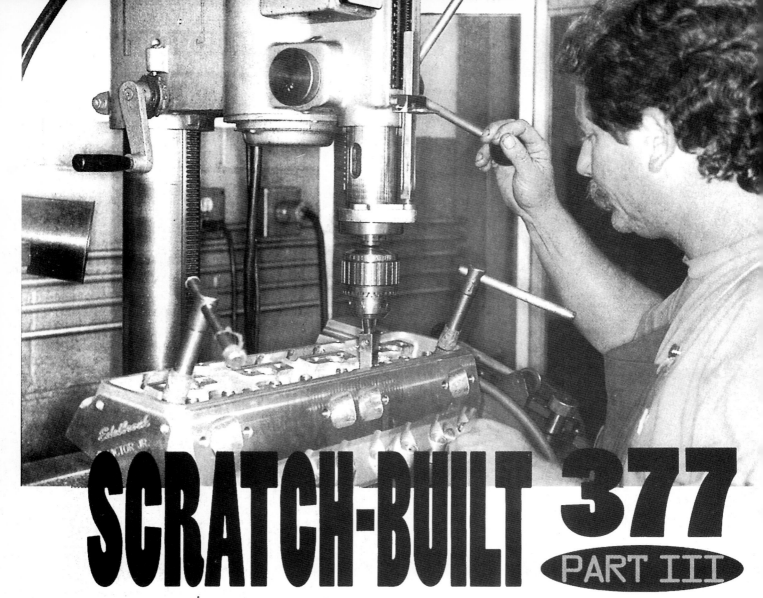

SCRATCH-BUILT 377
PART III

A SOON-TO-BE "HOT STREET" ENGINE GETS CYLINDER HEADS AND FINISHING WORK ON THE BLOCK AT SHAVER RACING

By Mike Magda

Our attention now turns to the cylinder heads. If you've been following this scratch-built 377 project in consecutive order, you know that we have gathered the parts necessary for Shaver Racing in Torrance, California, to build a 377ci high-strung racing small-block. A 377 engine is achieved by running a 3.48-inch-stroke crankshaft in a .030-over 400 block (final bore, 4.155 inches).

Originally, the purpose of this project was to show what's necessary to build a high-performance or race engine from scratch. We contacted experts to help in making all the parts decisions (see "Scratch-Built 377, Part I"), and then Shaver Racing started work on the block (see "Scratch-Built 377, Part II"). Now we're going to follow the final block prep, make some preliminary measurements to check clearances, and assemble the cylinder heads.

Soon, the engine's purpose will be fully defined when Shaver Racing wraps up final assem-

1 Following the machine work, the GM Performance cast-iron Bow Tie block was fitted with Pioneer soft plugs and painted black with VHT paint. This siamese-bore racing Bow Tie block comes CNC machined for splayed four-bolt main caps.

bly, followed by dyno testing. We are now building a "Hot Street" engine. The rule book for the *Hot Rod* Magazine World's Fastest Street Car Drag Racing Series has a class called Hot Street. It is designed for cars equipped with small-block engines, a single four-barrel, and no nitrous. There are no internal restrictions in the rules, and three-stage dry sumps are allowed. Commercially available cast-aluminum intake manifolds are required.

The Shaver scratch-built 377 fits perfectly into this class. We wanted a full-bore racing engine for possible bracket use when the project was started. With the exception of the dry-sump possibility, this engine should be competitive in this new class. Cars will run 9 pounds per cubic inch, so all that's needed is a 3,393-pound Camaro or Nova (100 pounds less if running an automatic transmission). Hot Street cars are restricted to 10.5-inch slicks.

The GM Performance Bow Tie block was machined in Part I, but some final prep work was needed. It was painted black with VHT paint, and the oil return holes were fitted with Moroso screens. Finally, Clevite 77 cam bearings were installed. Craig Johnson of Shaver Racing mocked up the short-block a few times to check clearances, such as crankshaft endplay and bearing clearances. He also assembled the Klein connecting rods and Wiseco pistons and installed the Speed-Pro rings in preparation for the final assembly.

2 Recapping some of the block modifications, Moroso deck plugs were installed, and the cylinders were bored out to accommodate 4.155-inch pistons.

5 The Bow Tie block is designed for crankshafts with two-piece rear main seals.

8 The bottom half of the rear main bearing was grooved slightly to allow oil relief to the front. This takes some of the pressure off the rear main seal.

3 A Moroso oil screen kit was installed over the oil return holes. The screens will prevent large pieces of a lifter from entering the oil pan should one shatter during operation. Oil drain-back holes over the camshaft were fitted with tubes to relieve pressure in the crankcase and prevent oil from interfering with the camshaft. Note that the lifter bosses were machined down to clear the links on the Isky roller lifters.

6 Craig Johnson of Shaver Racing handled the preassembly to check clearances. The top halves of the main bearings were installed, and the GM Performance steel crank was positioned to check for endplay. Johnson looks for .004- to .010-inch play. He also checks when the main cap is installed, since the clearance could tighten up slightly. Note that Johnson doesn't mount the indicator holder directly to the block, but instead uses a machined plate as a consistent, solid foundation.

9 The Wiseco pistons were checked one final time before being fitted to the Klein connecting rods and wrapped with Speed-Pro rings.

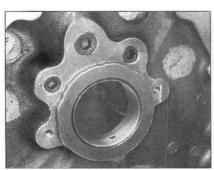

4 Clevite 77 camshaft bearings were installed. Note how the cam bearing boss was machined down to clear the timing chain.

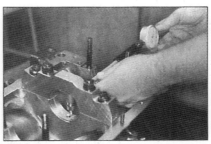

7 Bearings were fitted into the 8620 steel main caps and were checked for clearance against crankshaft measurements. Rod bearings were also checked. Clearance came in at .0025 to .0026 inch for all measurements.

10 Final piston-to-cylinder-wall clearance was .0045 inch.

SCRATCH-BUILT 377, Part III

While Johnson was finishing up the block, Steve Selotti worked on the Edelbrock Victor Jr. cylinder heads. The heads come race-ready with extensive CNC machining in the raised exhaust ports and combustion chambers. Intake runner volume is a generous 212 cc.

The heads are shipped with 70cc combustion chambers, which wouldn't offer much compression even with the Wiseco pop-up pistons. Selotti shaved .036 inch from the head surface to reduce the chambers to 64 cc. Final cc'ing of the heads and cylinders with pistons revealed a final compression ratio of 13.5:1, which may be a little low for an all-out screamer but gives us a good starting point.

The heads come with 2.08/1.60 stainless steel valves, but the rest of the valvetrain comes from Isky Racing Cams. Isky recommended upgrading to 1.560-inch-diameter valvesprings, so the spring pockets were cut out on the Edelbrock heads. Selotti also checked the spring pressure, installed spring height, and valve depth to ensure consistency throughout the heads. Before final assembly, the valves were marked, deburred, and lapped.

Keep reading. In the next story in this series, part four, we'll follow the final assembly and dyno testing at Shaver Racing. Barry Grant

11 The Klein Diatron connecting rods were fitted to the Wiseco pistons with free-floating pins with spiral locks in preparation for final assembly.

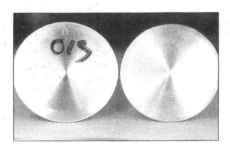

12 To ensure enough valve-to-piston clearance, the valve faces were taken down .015 inch on a lathe. Note the difference between an unfinished valve and a finished one.

13 Steve Selotti of Shaver Racing deburred the valves before installation. A grooved rubber wheel worked on the stems.

14 The other end got treatment on a Scotch-Brite wheel. Deburring makes for easier installation into the guides and through the seals.

15 Larger valvesprings were necessary on the Edelbrock heads, so Selotti first checked the amount of metal under the spring pocket before machining. An absolute minimum of .060 inch is needed, but the Edelbrock had upward of .300 inch of available metal before hitting the water jacket.

16 The spring pockets were machined out to 1.66 inches. Note the difference between the machined pocket and the stock pocket on the right.

17 The valve job was checked with a depth gauge. The gauge is zeroed on the deck, and then the depth of the valves is checked on the edge. All valve depths should be equal.

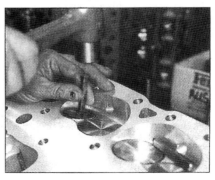

18 Before lapping the valves in place, Selotti numbered each valve according to the cylinder.

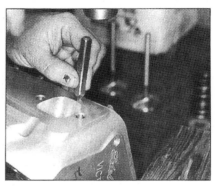

19 For easy reference, he also marked the cylinder heads just above the exhaust port.

has supplied a variety of carbs for testing, and we're looking to pull more than 600 hp. Using the Mr. Gasket Desktop Dyno software, we came up with 609 hp at 7,500 rpm with peak torque of 514 lb-ft at 5,500 rpm. We can't wait for the real test. ■

20 The valves were lapped in place.

21 The Isky valvesprings were checked at 250 pounds at 1.970 inches.

22 The installed height was also checked at 1.970 inches.

23 The heads were fitted with Teflon valve seals.

24 The valves and springs were installed with Isky 10-degree titanium retainers and locks. The heads were finished, with the exception of the No. 1 cylinder. The unfinished head was delivered to Johnson in the engine room so he could check valve-to-piston clearance. Once Johnson obtained the possible valve movement, the head was returned for final assembly.

25 Here's a sneak preview of what's to come in the rest of the series. The short-block is now assembled and the heads installed, so we'll have the complete assembly and dyno testing in the next installment.

SOURCES

Clevite 77
Dept. CHP
325 E. Eisenhower Pkwy., Ste. 202
Ann Arbor, MI 48108
313/663-6400

Edelbrock
Dept. CHP
2700 California St.
Torrance, CA 90503
310/781-2222

GM Performance
See your local GM dealer or call 800/577-6888

Isky Racing Cams
Dept. CHP
16020 S. Broadway
Gardena, CA 90247
213/770-0930

Klein Engines
Dept. CHP
1207 N. Miller Rd.
Tempe, AZ 85281
602/967-5990

Moroso
Dept. CHP
80 Carter Dr.
Guilford, CT 06437
203/453-6571

Pioneer Automotive
Dept. CHP
5184 Pioneer Rd.
Meridian, MS 39301
601/483-5211

Shaver Specialty
Dept. CHP
20608 Earl St.
Torrance, CA 90503
310/370-6941

VHT
Dept. CHP
8747 E. Via del Commercio
Scottsdale, AZ 85258
602/991-8002

Wiseco Pistons
Dept. CHP
7201 Industrial Park Blvd.
Mentor, OH 44060
800/321-1364

HOT STREET

THE SHAVER RACING SCRATCH-BUILT 377 FITS PERFECTLY INTO THE NEW *HOT ROD* DRAG RACING SERIES

By Mike Magda

The last time we checked, the rule book for the *Hot Rod* Magazine World's Fastest Street Car Drag Racing series was sold out. One big reason is the enthusiasm for the Hot Street class, which is for vehicles powered by small-block engines no larger than 385 ci. When we originally started the Scratch-Built 377 project, the rule book hadn't been released. Our original plan in following the project at Shaver Racing was to detail an engine buildup starting from scratch based on a challenge to "build a 377 engine." From there, we contacted industry experts and gathered parts to build a full-race version of the big-bore, short-stroke combination. The new rule book was released just as we finished work on the GM Performance Bow Tie block and started prepping the Edelbrock Victor Jr. heads. We noticed that the Shaver engine would fit nicely into the class, since it was designed from the beginning to be a high-end bracket racing engine. Now, however, we have a Hot Street engine.

The Hot Street rules allow any internal modifications but prohibit supercharging, turbocharging, and nitrous oxide. Any aftermarket cylinder head is allowed, but induction is restricted to cast manifolds and a single four-barrel carburetor. Since one of our original intents with this engine project was to show how to build an engine from scratch, we restricted our parts selection to off-the-shelf products. Dry-sump oiling is allowed, but we held back at this point. A possible future update could include the installation of a dry-sump system to judge the difference in horsepower. The Bow Tie block is dry-sump ready, but to take full advantage of the dry sump we would have to change pistons and rings.

If you've been following this project in sequence, you know that Shaver Racing prepped the Bow Tie block issue by milling the lifter bosses to clear the roller lifters; shaved the deck for an 8.990 deck height; bored and honed the cylinders to 4.155 inches; align-honed the main bearings; and installed a Moroso screen kit and deck plugs, Pioneer freeze plugs, and an oil restrictor kit.

Once that was done, the Edelbrock heads were shaved to boost the compression ratio and were fitted with larger Isky springs. Much of the pre-assembly test-fitting and measuring took place so that the final assembly would not be hampered.

We follow Craig Johnson of Shaver Racing as he assembles the engine in this issue, and we'll dyno test it in the next segment. Then all we'll need is a Hot Street car for track testing. Stay tuned.

1 The Wiseco pistons with a 12cc dome and 1.250 compression height were fitted with Speed-Pro plasma-moly rings.

2 Clevite 77 bearings were installed in the block and main caps, and then the GM Performance 3.48-inch-stroke crankshaft was positioned in the block. A dab of clear silicone (arrow) was placed on each side of the block for sealing before the rear main cap was installed.

3 Craig Johnson torqued down just the rear main cap and checked the crankshaft endplay before installing and torquing down the intermediate caps to 65 lb-ft. Then the endplay was checked again.

4 The Isky roller cam was installed, followed by an adjustable Moroso double-roller timing set.

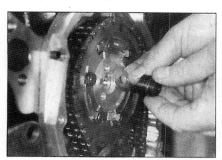

5 The timing set comes with a cam button to limit camshaft endplay. The amount of play is determined by the shims provided in the kit.

6 A Manley three-piece timing cover was installed. The top half must be in place to hold the cam button. Johnson set the endplay at 0.001 to 0.005 inch.

7 The Wiseco piston/Klein Diatron rod assemblies were installed with the help of Powerhouse ring compressors.

8 Once all the pistons were installed, the coolant holes on the decks were prepped with clear silicone before the Fel-Pro head gaskets were positioned.

9 The Edelbrock heads were positioned and snugged in place with just a couple of bolts.

10 Valvesprings on the No. 1 cylinder were not installed so valve-to-piston clearance could be checked. The piston was brought to TDC, and valve travel was gauged with a dial indicator. The head was removed and valve-springs installed. When the cam timing is checked, the actual valve lift at TDC will be measured and subtracted from the full-travel measurement to determine valve-to-piston clearance.

11 When all the clearances were checked, the heads were bolted in place. We used ARP head bolts. Johnson lubricated the bolt thread and both sides of the washer before torquing the heads down to 67 lb-ft.

12 The Isky roller lifters and 0.100-inch-longer pushrods were installed.

13 ARP 7/16-inch rocker studs were installed. Some studs intruded on the intake ports and had to be sealed with silicone to avoid air leaks.

HOT STREET

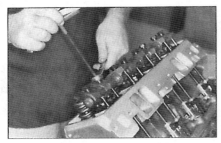

14 The Isky roller rockers were installed. Intakes were treated to 1.65:1 rockers, while the exhaust got 1.5:1 units. Total valve lift was 0.714 inch on the intake and 0.645 inch on the exhaust. Cam duration was 264/272 at 0.050, and the lobe centers were 108 degrees apart.

15 The Moroso adjustable cam set allows for quick changes in cam timing. The Isky cam was installed at plus 2 degrees advanced.

16 After all the milling on the deck, heads, and manifold, extra-thick Fel-Pro intake gaskets were needed. These are trim-to-fit gaskets, so Johnson carved out the openings on each one.

17 The Edelbrock Victor Jr. intake manifold was bolted down with ARP bolts. An inspection light helped check the port alignment.

18 The Moroso oil pump was bolted in place and safety-wired.

19 The Moroso Stage II Pro Eliminator 6-quart pan was a little tight around the mains, so the areas were marked and ground down with a die grinder for a smooth fit.

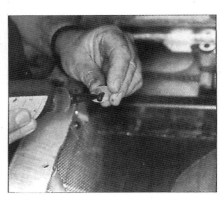

20 The windage-tray screen was secured with Loctite-treated bolts before final installation.

21 The Fel-Pro gaskets and seals were secured with gasket adhesive, and clear silicone was applied on the seals and around the top of the gasket. The Moroso only has three side mounting locations, so every precaution against leaks was taken.

22 The oil pan was bolted in place. Again, more silicone was used to seal up the mounting locations.

23 A GM Performance mini starter was installed.

24 The Manley timing cover and a TCI Rattler harmonic balancer were installed. The balancer was secured with an ARP bolt.

25 A Moroso electric water pump was installed.

26 Spark will come from a Mallory Comp 9000 billet breakerless distributor hooked up to a HyFire IV capacitive-discharge ignition box and Promaster coil. Mallory ignition wires were also used. The distributor, which has to be removed for oil priming, will be held in place with a Mr. Gasket clamp.

27 BG Fuel Systems supplied a variety of carbs for testing, including 650- and 750-cfm models.

28 The GM Performance valve covers were painted black and fitted with K&N breathers.

29 The final dress-up included a water neck and fuel pump block-off plate from TD Performance. We're looking for 600 hp, so check out "Hot Street Power Test" for a full report on the dyno tests. ■

SOURCES

ARP
Dept. CHP
531 Spectrum Cir.
Oxnard, CA 93030
805/278-7223

BG Fuel Systems
Dept. CHP
1450 McDonald Rd.
Rte. 1, Box 1900
Dahlonega, GA 30533
706/864-8544

Clevite 77
Dept. CHP
325 E. Eisenhower Pkwy., Ste. 202
Ann Arbor, MI 48108
313/663-6400

Edelbrock
Dept. CHP
2700 California St.
Torrance, CA 90503
310/782-2900

Fel-Pro Performance
Dept. CHP
7450 N. McCormick Blvd.
Skokie, IL 60076
708/568-2376

GM Performance
See your local General Motors dealer or call 800/577-6888

Isky Racing Cams
Dept. CHP
16020 S. Broadway
Gardena, CA 90247
213/770-0930

K&N Engineering
Dept. CHP
561 Iowa Ave.
Riverside, CA 92502
909/684-9762

Klein Engines
Dept. CHP
1207 N. Miller Rd.
Tempe, AZ 85281
602/967-5990

Mallory Ignition
Dept. CHP
550 Mallory Way
Carson City, NV 89701
702/882-6600

Manley Performance
Dept. CHP
1960 Swarthmore Ave.
Lakewood, NJ 08701
908/905-3366

Moroso
Dept. CHP
80 Carter Dr.
Guilford, CT 06437
203/453-6571

Mr. Gasket
Dept. CHP
8700 Brookpark Rd.
Cleveland, OH 44129
216/398-8300

Pioneer Automotive
Dept. CHP
5184 Pioneer Rd.
Meridian, MS 39301
601/483-5211

Powerhouse Products
Dept. CHP
3402 Democrat Rd.
Memphis, TN 38118
901/795-7600

Shaver Specialties
Dept. CHP
20608 Earl St.
Torrance, CA 90503
310/370-6941

TD Performance
Dept. CHP
16410 Manning Way
Cerritos, CA 90703
562/921-0404

Wiseco Pistons
Dept. CHP
7201 Industrial Park Blvd.
Mentor, OH 44060
800/321-1364

HOT STREET POWER TEST

WE JUST MISS 600 HP WITH THE SHAVER RACING SCRATCH-BUILT 377

By Mike Magda

PHOTOGRAPHY: MIKE MAGDA

The goal with our scratch-built small-block was to pull 600 hp or more from a 377ci engine using off-the-shelf parts with minimal machining. We almost made it, but before we disclose the final dyno results, let's recap.

We originally started this project as an exercise to show what it takes in parts and manpower to build an engine from scratch. The 377ci displacement was chosen following a challenge from a reader who says all we work on are 355s and 383s. When we started to gather the parts, the new *Hot Rod* Magazine World's Fastest Street Car Drag Racing Series rule book came out, announcing a new class called Hot Street. This class was designed for cars with small-blocks no larger than 385 ci, single four-barrels, and no nitrous. Our 377 seemed to fit, so we opened up our options as far as they could go with the parts we had already ordered.

The engine was mounted on Shaver Racing's Heenan-Froude dyno. Exhaust was scavenged through a set of 1⅞-inch dyno headers.

We found a home base for this project at Shaver Racing in Torrance, California. Known for building winning engines for drag racing, Sprint cars, and off-road racers, Shaver Racing opened up its machine shop and assembly rooms to our cameras as the project unfolded.

We started by gathering all the parts and tools necessary for the job, to prevent any delays during the buildup. The GM Performance engine block was prepped by machinist Ray Bates at Shaver Racing. Some of the work performed included machining down the cam bearing boss to clear the timing chain, boring and honing the cylinders out to 4.155 inches, clearancing the lifter bosses for roller lifters, installing deck plugs, and balancing the rotating assembly. Steve Selotti prepped the Edelbrock Victor Jr. cylinder heads in the March issue by blueprinting the valves, enlarging the spring pockets, setting the installed spring heights, and milling the head surfaces for a final 64cc combustion chamber.

All during the machine work and cylinder-head preparation, Craig

Johnson took measurements and pre-fitted the parts to ensure proper clearances. The final assembly has already been covered.

The engine was mounted on a Heenan-Froude dynamometer with DEPAC data acquisition. A set of 1⅞-inch dyno headers was installed. Johnson primed the engine with 30W racing oil before startup. The 377 was broken in with a 45-minute run at around 3,000 rpm. Johnson inspected the engine for leaks and checked the valves before Ed Hansen came in to run the dyno tests. A few warm-up runs were made to check exhaust-gas temperatures and brake-specific fuel consumption and to make final timing adjustments. The first test was made with a Holley 750 prepped by BG Fuel Systems. Modifications included custom-finished venturis, removed choke horns, dual-feed double-pumper bowls, and four-corner idle. Right out of the box, there were no hot spots in the exhaust temperatures or an indication of lean brake-specifics, so the jets were not changed.

The best pull with the 750 carb produced a peak of 583 hp at 6,500 rpm with peak torque of 514 lb-ft at 5,100 rpm. A Barry Grant 850 carb with annular boosters was installed for the next test. The bigger carb raised the peak power level to 6,700 rpm, with 588 hp being registered. Peak torque stayed the same. No jet changes were made due to time constraints, but the brake-specific numbers followed closely with the first test. For the final run, a 2-inch TD Performance spacer was installed between the intake manifold and carburetor. Peak power crept up a little to 589, but torque went up to almost 519 lb-ft at 5,300 rpm.

With a little more time, we probably could have found 11 more hp, but there was a long line of racing engines waiting to get on the dyno in preparation for the upcoming season. We could have retarded the cam a little and played with the carbs and timing to stretch the numbers up.

Would this engine be competitive in Hot Street? It might be a qualifier in the first year, but the experts we consulted agreed that 700 hp would be needed to win. How could we pick up another 100 horses? First, we would have to go up to a 14.5:1 or 15:1 compression ratio. The rules allow for dry-sump oiling, so that would help quite a bit. The next big jump could come with fully modified 18-degree heads and a matching intake manifold. A slightly bigger camshaft should put us over the top, but then you're looking at another $5,000 to $7,000 worth of improvements. For an off-the-shelf engine, this 377 is quite potent, with plenty of room for improvement.

Ed Hansen handled the dyno runs. A few preliminary pulls indicated that 35 total degrees of timing worked the best.

Shaver Racing's dyno recorded a peak of 589 hp on the final run.

The first test was run with a Barry Grant 750, and then a switch was made to a BG 850 with annular boosters.

For the final test, a TD Performance 2-inch spacer was installed.

HOT STREET POWER TEST

INSIDE THE HOT STREET 377

The Hot Street 377 is based on a GM Performance Bow Tie block bored out to 4.155 inches. It was fitted with Clevite 77 bearings, Pioneer soft plugs, and assorted hardware from Moroso, Mr. Gasket, and TD Performance. The rotating assembly consists of a GM Performance 1053 3.48-inch steel crankshaft, Klein Diatron 6-inch rods, Wiseco pop-up pistons, Speed-Pro rings, a TCI Rattler harmonic balancer, a B&M flexplate, and assorted hardware from Pioneer and ARP. The valvetrain is all Isky, starting with a 264/272-duration roller camshaft, 0.100-inch-extra-long pushrods, heavy-duty (230 pounds at 1.970 inches) springs, and titanium retainers and valve locks, plus 1.65:1 intake rockers for 0.714 lift at the intake valve and 1.5:1 exhaust rockers for 0.645 at the exhaust. The cam is turned by an adjustable Manley roller timing chain and is covered by a two-piece Manley cover.

The oiling is handled by a complete system from Moroso that includes a Stage II Pro Eliminator 6-quart pan, a high-performance oil pump, and assorted hardware. Up top, the Edelbrock heads come bare with stainless steel valves. Shaver added ARP 7/16-inch rocker studs. Induction is via carburetors from BG Fuel Systems and an Edelbrock Victor Jr. intake manifold. Spark is provided by a Mallory billet distributor and wires.

Other engine parts include Fel-Pro gaskets, Russell plumbing, a Moroso electric water pump, K&N breathers, Champion spark plugs, a GM Performance starter, and dress-up goodies from GM Performance and TD Performance. Some engine-building tools were supplied by Pioneer, Moroso, Eastwood Company, JAZ Products, and Powerhouse.

Special thanks to Ron Shaver and Lee Brewer and the rest of the machinists and engine builders at Shaver Racing. ■

The bottom end is all Moroso, while the top end is Edelbrock and fed by BG Fuel Systems.

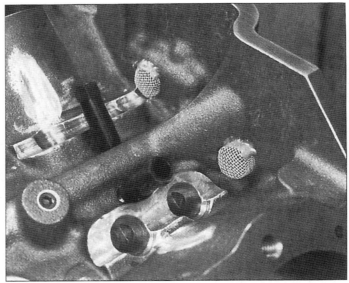

The GM Performance Bow Tie block was race-prepped with debris screens, crankcase vent tubes, and significant machining to clear the roller lifters, timing chain, and connecting rods.

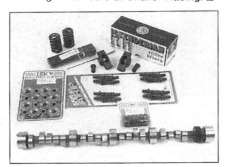

The valvetrain was supplied by Isky and included the camshaft, roller lifters, pushrods, valvesprings, retainers, locks, and roller rockers.

The rotating assembly consisted of a GM Performance crankshaft, Klein rods, and Wiseco pistons.

DYNO RESULTS

	TEST #1			TEST #2			TEST #3		
RPM	TORQUE lb-ft	HORSE-POWER	BSFC #/HrHP	TORQUE lb-ft	HORSE-POWER	BSFC #/HrHP	TORQUE lb-ft	HORSE-POWER	BSFC #/HrHP
4,200	422.1	337.6	.40	437.2	349.5	.41	434.4	347.5	.41
4,400	450.7	377.6	.36	458.1	383.8	.39	454.2	380.6	.39
4,600	482.1	422.2	.33	486.5	426.1	.37	489.7	429.0	.35
4,800	502.0	458.7	.33	504.9	461.5	.36	507.7	463.9	.36
5,000	**511.9**	487.3	.34	512.8	488.2	.37	513.9	489.3	.39
5,200	511.8	506.7	.36	**513.0**	507.9	.39	515.8	510.6	.40
5,400	511.6	526.0	.39	511.3	525.6	.41	**516.1**	530.6	.41
5,600	506.7	540.3	.39	509.8	543.5	.41	513.7	547.7	.39
5,800	501.7	553.9	.40	504.5	557.2	.41	508.5	561.5	.41
6,000	497.2	568.0	.40	500.4	571.7	.41	501.6	573.0	.42
6,200	488.1	576.2	.41	491.0	579.7	.41	494.2	583.3	.42
6,400	478.0	582.4	.42	479.4	584.2	.43	483.3	589.0	.43
6,500	471.3	**583.2**	.42	472.3	584.6	.44	476.0	**589.2**	.44
6,600	463.7	582.8	.42	465.7	585.1	.44	467.5	587.4	.45
6,700	456.8	582.8	.43	461.2	**588.3**	.44	459.9	586.7	.45
6,800	449.4	581.8	.43	451.8	585.0	.45	452.2	585.5	.45
6,900	439.2	577.0	.44	442.3	581.2	.45	443.7	582.9	.45
7,000	427.5	569.7	.45	427.7	570.1	.46	432.5	576.4	.46

*Numbers in bold type denote peak readings.

THE GREAT 408

Build A 600hp Race Small-Block Using Lunati's Line Of Mail-Order Parts

Bracket racing has become so sophisticated in the last few years that it seems as if there's more emphasis on installing the latest high-tech electronic gizmos to improve reaction times than there is on building a potent engine/chassis combo, which is what actually moves you down the dragstrip in short order. Savvy car crafters know that it takes reliable equipment to consistently go fast, but everybody seems to overlook the fact that you can still build a killer race-engine combo without breaking the bank—if you choose the right combination.

Chris Padgitt switched from a door-slammer race car to a dragster a short time ago so he could run quicker without all of the hassles involved with running a heavy chassis. Initially, the dragster used Mopar power, but recently Chris decided to put together a small-block Chevrolet V-8 combo that he had recommended to many of his mail-order customers at Lunati Cams. Chris is a technical advisor at Lunati, where he works closely with street machiners and racers running NHRA and IHRA Sportsman classes, as well as weekend bracket racers. Pretty stout credentials, we'd say.

Chris' engine buildup goal was two-fold: to build a race 408

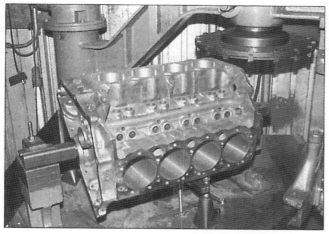
A fully machined Lunati Race Ready block is bored and honed with a torque plate, decked, align-bored, clearanced for a stroker crank, and Magnafluxed. All bolt holes are retapped and chamfered. Camshaft bearings are also installed.

The Lunati block comes clearanced for use with a stroker crankshaft. The cylinder bores are notched (arrows) to accept up to a 4.00-inch-stroke crank in a 400 block and up to a 3.75-inch-stroke crank in a 350 block.

PHOTOS BY JOHN KIEWICZ

The Lunati Race Ready block includes oil-restrictor plugs (A) for use with a roller camshaft and oil screens epoxied to the block (B). The lifter valley and timing chain valley are covered with high-temp paint (C) to aid oil flowback, and the outside of the block is painted black (D).

The Lunati Pro Series cranks use a Contoured Wing design, which incorporates specially shaped counterweights. This design allows the rushing air and oil to be directed to the main bearings rather than onto the connecting rods. The result is better oiling, less enertia loss from windage, and more horsepower.

The Lunati Pro cranks are constructed of 4340 nontwist steel. To reduce reciprocating weight, all rod journals are drilled with either a ¼- or ⅜-inch lightening hole.

The Great 408

that delivered 600 hp for less than $5,000 and to use off-the-shelf parts that could be assembled by most any amateur engine builder. It's true that you can build an engine using reworked stock parts that will turn out about 500 hp, but if you regularly race the engine, it just won't live like an engine built with high-quality performance parts. Thus, to build an engine that will survive countless high-rpm blasts, you're going to need to spend some money on good aftermarket parts such as a heavy-duty crank, rods, and pistons, as well as high-flow heads. All of these parts and more are available directly from Lunati, which makes the company a perfect one-stop source for a race-engine buildup.

The Bottom End

The core component for Chris' 408 bracket-race engine was a fully reworked cylinder block from Lunati. A Lunati Race Ready block is fully machined to race specs, includes steel four-bolt main caps, and is clearanced for a stroker crankshaft.

A Lunati Pro Series 4340 (3.75-inch-stroke) crankshaft was chosen to handle the high-horsepower, high-rpm abuse the engine would encounter. The Pro Series cranks incorporate Lunati's new Contoured Wing counterweight design, which is used to move oil away from the rotating mass and onto the main bearing caps. This, in turn, delivers improved horsepower output. The crank uses a nontwist forging and is precision-built, thus requiring little or no heavy metal to balance.

Good connecting rods are a must for high-compression engines that operate at high rpm. That's why Chris chose Lunati Pro Mod rods that are forged from 4340E alloy steel. The rods are profiled to provide a combination of outstanding strength and light weight yet still remain affordable. The Lunati rod bolts provide a 220,000-psi clamping force, which ensures long life under the stresses of all-out racing. To improve the rod ratio (which reduces cylinder side loading and increases torque), 6-inch center-to-center-length Lunati Pro Mod rods were used in place of stock-length (5.565- or 5.7-inch) rods.

To build big power, a healthy compression ratio is needed. Thus, Chris opted for Lunati & Taylor 0.040-inch overbore pistons that deliver a 13.3:1 compression ratio with a 0.250-inch dome. The L&T pistons are constructed of forged aluminum and are machined to stringent tolerances, which delivers optimum performance along with a long life. Lunati Pro Series plasma-moly rings team up with the L&T pistons to properly seal the high combustion pressures.

Each Lunati Pro Series crank has journal radii that are ground to 0.125 inch. The roundness of each journal is within 0.0001 inch or less, and overall stroke is held to plus or minus 0.001 inch. The surface finish of the shaft is a Grade 5 RMS or better, and each shaft is heat-treated by plasma gas nitrite, which creates a case depth of 0.015-0.022 inch.

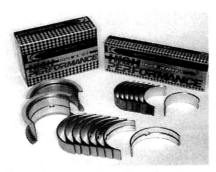

The Lunati 408 was fitted with King Pro Alecular bearings that are constructed of alloy, tin, and copper. The bearings feature a surface depth of 0.012-0.015 inch as compared to most other bearings that have a thin babbitt overlay of 0.0005-0.0008 inch. This increased surface depth provides added strength and resists bearing journal scuffing.

The Top End

With a bulletproof, high-performance bottom end built, a high-flowing top end was needed. Regulating intake airflow is a Lunati roller camshaft actuating mechanical lifters. Cam specs are healthy in the form of a 0.734-inch-lift intake and a 0.666-inch-lift exhaust, a 282-degree-duration intake and a 292-degree-duration exhaust, and a 113-degree lobe separation with a 109-degree intake lobe centerline. Lunati Pro Series pushrods bridge the gap between the roller lifters and Lunati 1.6:1 (intake) and 1.5:1 (exhaust) roller-rocker arms. A Lunati aluminum rocker-arm stud girdle was used to prevent unwanted valvetrain flex at high rpm.

To shuttle gobs of airflow inward, Brodix Track I aluminum cylinder heads were used, providing ample breathing for the 408 small-block. The Brodix heads feature 69cc combustion chambers with 2.08-inch intake and 1.60-inch exhaust valves that provide sufficient airflow for the 408. The heads were gasket-matched and had mild blending work done on the bowls, but otherwise they were stock out of the box. The heads were capped with Brodix aluminum

Lunati Pro Mod rods combine outstanding strength with a real-world price tag, thus making them perfect for a bracket race engine. The rods are custom-profiled on a CNC machine, and the I-beams are polished to help roll oil away from the rods. The rods' small ends are fitted with Amco 18 bronze bushings, and the big ends are fitted with ARP 7/16-inch rod bolts. The Pro Mod rods are weight-matched to plus or minus 1.5 grams per end.

To build big power, the 408 was fitted with Lunati & Taylor lightweight pistons that netted a 13.3:1 compression ratio. The L&T pistons are forged from high-silicone-content 4032 aluminum and are machined to extremely tight specifications to withstand the abuse of racing conditions.

Lunati 408 Parts List

PN	Item
ARK23111	Lunati Race Ready Engine Block
AJ-211	Lunati Pro Crank (3.750-inch stroke)
LAD-1	Lunati Pro Mod Rods (6.00-inch length)
1412B2S5	Lunati & Taylor forged pistons (4.165-inch bore, 0.250-inch dome, 1.120 comp. height)
P14170	Lunati Pro Series Plasma Moly ring set
CR848HP	King Alecular rod bearings
MB5143HP	King Alecular main bearings
93100-9	Lunati nine-keyway timing chain set
50199	Lunati steel billet roller camshaft
72840	Lunati mechanical roller lifters
84156	Lunati roller-rocker arms
91644	Lunati stud girdle with adjusting nuts
82141	Lunati Pro Series chrome-moly pushrods
Track 1	Brodix Track 1 aluminum cylinder heads
Weiand 7531	Weiand Team G aluminum intake manifold
SBC	Brodix aluminum valve covers
HEI Race	Performance Distributors HEI distributor
850 cfm	DynoTech 850-cfm Holley carburetor
91708	ATI Torsional Super Damper

valve covers.

A Weiand Team G aluminum intake manifold and Dyno-Tech 850-cfm Holley double-pumper carburetor were used to funnel the air/fuel mixture inward. ARP bolts were used to provide tight clamping forces, and Fel-Pro gaskets were used to ensure a leak-free seal. To provide ample spark, Chris installed a Performance Distributors (PD) HEI distributor, PD Live Wires, and a PD Terrorist 18-volt battery.

Brodix Track I heads are proven performers, so a set was installed on the 408. Standard features include special intake and exhaust runners matched with 2.08-inch intake and 1.60-inch exhaust valves. A somewhat large, 69cc combustion chamber helps unshroud the valves, thus further improving airflow. The head's aluminum construction sheds substantial weight off the engine and also dissipates combustion heat better, thus thwarting detonation.

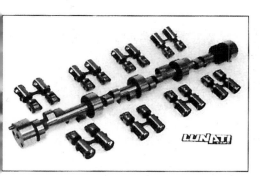

A rowdy mechanical roller camshaft from Lunati was used in the 408. The cam, constructed of billet steel, makes use of 0.734-inch-lift (intake) and 0.666-inch-lift (exhaust) specs to deliver plenty of air/fuel mixture. Special lightweight, heat-treated roller lifters from Lunati help reduce reciprocating mass for improved revving and reduced valvetrain stress.

The 408 was fitted with a Lunati nine-keyway roller timing chain set that is constructed of chrome-moly steel. The crank gear has nine slots machined into it, thus allowing the camshaft to be advanced or retarded 1, 2, 3, or 4 degrees. The timing gear set also allows for the cam to be installed straight up if desired.

Heavy-duty valvesprings, retainers, and keepers work together to prevent valve float with high-lift camshafts. ARP screw-in studs and Lunati guideplates help deter valvetrain flex and parts failure.

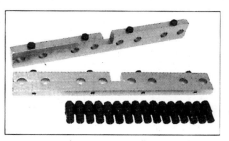

A Lunati stud girdle was fitted to the 408 to prevent valvetrain flex during high-rpm operation. The girdle is constructed of aluminum and is a simple bolt-on. The Lunati stud girdle includes fully adjustable, locking rocker-arm nuts.

Testing

Since all of the Lunati parts were precision-machined and ready to install right out of the box, Chris went to work building the engine using traditional engine-building tools. After being built, the completed Lunati 408 was taken to Dyno-Tech in Memphis, Tennessee, for break-in and fine tuning on an engine dyno. Afterward, the engine was flogged at all rpm levels to determine its horsepower output. Power output was a potent 591.3 hp at 6,750 rpm, along with 515.5 lb-ft of torque at 5,500 rpm. ■

The Brodix Track I heads were fitted with Lunati roller-rocker arms constructed of aluminum. The rocker arms use a roller tip along with a needle-bearing rollerized center to reduce valvetrain friction. For improved horsepower, 1.6:1 rockers were used on the intake side, and stock-ratio 1.5:1 rockers were used on the exhaust side.

Lunati Pro Series pushrods were used to both handle the rigors of high-rpm abuse as well as to contend with the high valvespring pressures needed with the high-lift racing camshaft. The pushrods are constructed of heat-treated chrome-moly steel and have specially formed one-piece swaged ends.

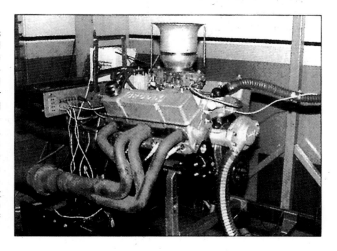

Lunati 408 Engine Test

RPM	Horsepower	Torque
4,500	388.6	453.5
4,750	400.1	442.4
5,000	451.4	474.1
5,250	498.7	498.9
5,500	539.8	**515.5**
5,750	559.6	511.1
6,000	566.4	495.8
6,250	571.1	479.9
6,500	585.0	472.7
6,750	**591.3**	460.1
7,000	582.0	436.7

Sources

Brodix Cylinder Heads
Dept. CC
P.O. Box 1347
Mena, AR 71953
501/394-1075

DynoTech
Dept. CC
5516 Old Hwy. 78
Memphis, TN 38118
901/795-3000

Lunati Cams Inc.
Dept. CC
P.O. Box 18021
Memphis, TN 38181
901/365-0950

Performance Distributors
Dept. CC
2699 Barris Dr.
Memphis, TN 38132
901/396-5782

Weiand Automotive Industries
Dept. CC
P.O. Box 65301
Los Angeles, CA 90065
213/225-4138

OUTER LIMITS SMALL BLOCK

Going To The Edge Of The Envelope With A 427 Small-Block
Part I—Machining The Block

By Jim Losee

Do not attempt to adjust your eyes, as we have taken control of them. You did read the title correctly. *Yes,* it says 427, and *yes,* it is a cast-iron *small*-block.

Coming up with the desired 427 cubes of cylinder displacement requires a bore of 4.125 inches in diameter and a stroke of 4.00 inches even. To get the rod-to-stroke ratio back to some semblance of normalcy, a connecting rod that is 0.150 inch longer than stock was required. This puts the center-to-center distance at 5.850 inches versus the stock 5.7 inches.

The purpose of building this big small-block is not only to show how it's done and what's involved, but also to demonstrate just how far you can go with the technology and the aftermarket parts available for the small-block Chevy of today. We are going to be testing the 427 with a carburetor as an induction source, and in a few months' time, when we're done with this buildup, we'll use this engine to run some tests on an electronic fuel injection unit from Street Legal Performance. This displacement size will stretch the bounds of airflow capacity in most induction systems and it will help to test the outer limits of camshafts and cylinder heads as well.

Getting this kind of displacement from a cast-iron small-block Chevy was almost unheard of in the past. Major machine work that included the use of special oversized sleeves, deck spacers to raise the deck height, and special pistons were just some of the things that were involved in making this many cubes from an iron small-block.

Chevrolet has solved the problem of making large-displacement small-blocks with the release of their Bow Tie cast-iron block with siamese bores (PN 10051183). Due to the extra-thick cylinder walls of this block, it will allow up to a 0.155-inch overbore, increased substantially from the stock 4.00-inch-diame-

Part of the standard procedure at Shaver Specialties is to deburr the iron Bow Tie block, both inside and out. Rounding sharp edges decreases the possibility that cracks will occur.

Small-Block Outer Limits

Not trusting the factory press-in cup plugs in the oil galleys at the front of the block, Shaver Specialties' Rich Troxall reams and taps the holes out to ⅜NPT and installs Allen plugs.

This photo dramatically shows the difference between where the two-bolt main cap sits and where the four-bolt cap does. The space between the bearing bore and the four-bolt cap is where the two-bolt cap would sit normally.

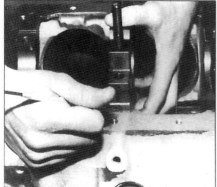

Rich is seen marking where the outer edge of the four-bolt cap meets the block. Screwing in the center ARP studs helps to locate the cap and when that is done the excess material is milled off so the four-bolt cap will sit flush.

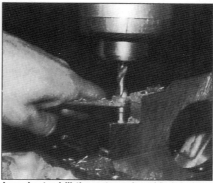

In order to drill the outer splayed bolt holes for the four-bolt caps, Shaver's uses a special fixture that aligns the milling machine head using a pilot. After the angle is determined with the pilot, the hole is drilled.

After the splayed outer cap hole is drilled, it is tapped. Again, Shaver's uses the milling machine to help guide the tap down the hole at the proper angle. When it's all done, it looks like a factory job.

Lubing the threads on the ARP main cap studs before they are tightened ensures that the correct torque will be achieved. The caps are torqued just prior to going into the horizontal align-boring machine.

ter bore of the standard block.

Other features of this block are a thicker deck surface than a standard-production block (to accommodate a better head gasket seal), blind headbolt holes that don't go into the water jackets, and main cap bulkheads which are thicker (to resist pulling out and to help give a better purchase when splayed outer-bolt, four-bolt main caps are used).

Some of the not-so-nice features of this block are that it comes in a two-bolt main cap configuration from the factory and it uses the '86-to-present one-piece rear main seal. These blocks are pretty rough castings that need a bit of massaging to clean them up, and all the machined dimensions of the block need to be checked for proper alignment and fit. Something else we learned was to check all the blind headbolt holes to be sure they're tapped and that a tap hasn't been broken off in the headbolt hole.

Because of the special machining requirements of the new Bow Tie blocks, we had to select an engine-machining facility that was familiar with these parts. One of the companies involved in doing development work on the Bow Tie block with Chevrolet is Shaver Specialties of Torrance, California. Ron Shaver

Small-Block Outer Limits

and his crew do more racing engine-building for Sprint cars and other forms of automotive competition than just about anyone in the world. They have quite a bit of experience making small-block Chevys live in the most severe racing environments.

When you step inside the Shaver facility, you know that this is what all engine-building shops ought to look like. Everything is clean and if it isn't new, it sure looks like it is. All the latest state-of-the-art engine-machining equipment is there, along with personnel who are very experienced.

We talked to Ron Shaver and his machinist, Rich Troxall, about the requirements for a street-driven 427 iron small-block. This dynamic duo figured that a street-driven 427 with a 6000-rpm redline would not require their full-on NASCAR block treatment. The one thing that really stands out in this project is the fact that converting from two-bolt main caps, to splayed outer-bolt four-bolt caps, is a very labor- and time-consuming proposition. Rich is such a fussy machinist that he checks everything two or three times to ensure the accuracy that Shaver Specialties engines are known for in the Sprint car world.

Right off the bat, the engine is deburred on the outside and inside to reduce the chance of stress cracks showing up. It sure makes the engine easier to handle when there aren't a bunch of sharp edges to deal with. Then the little things are done to the block to ensure its longevity: reaming and tapping the front oil galley holes to a ⅜-inch pipe thread and making sure there's enough clearance for our Summers gear drive setup. Then the rear oil galley plugs are subjected to the same procedures.

After the block had been deburred and the screw-in oil galley plugs installed, it was time for Rich to put the block on the Rottler boring machine. But just as the block was being picked up to be put on the machine, we noticed that one of the blind head bolt holes had a broken tap in it. This required sending the block to Tap-Ex in Gardena to have the tap removed.

Tap-Ex removes broken taps and drill bits by electronic disintegration, normally using a copper rod. After one attempt at using the copper rod had failed, Jim at Tap-Ex determined that the factory had used a pure carbide tap and it would take a special molybdenum rod to disintegrate the tap. When it was all done, the tap was gone and the threads in the hole were in perfect shape. It sure was easier than using an E-Z Out!

After the block came back to Shaver's, Rich put it on the boring machine. Because we were going from the stock 4.00-inch bore to a 4.125-inch bore, Rich removed 0.050 inch on two successive passes down each bore for a total of 0.100 inch removed. The last pass on the

Most of us assume that converting from two-bolt main caps to four-bolt caps is an easy machining process. After watching what it takes to accomplish this task, we've learned it isn't! The block must be leveled on the align-boring machine in order to guarantee a straight cut.

Each main cap must be cut three times, with each pass getting successively bigger. In between each pass, a dial-bore gauge is used to check that the proper amount of material has been removed and that the bearing bores are round.

boring machine removed 0.020 inch, for a total of 0.120 inch removed from each bore. The final 0.005 inch of material to be removed would come out on the final honing with a torque plate tightened down to the deck surface with 50 ft-lbs of torque and oil applied to each head bolt as lube.

With the 4.00-inch-stroke Crower forged crank passing the standard Magnafluxing and straightness check that Shaver's gives all cranks, Rich then dropped it into the block to check counterweight interference. The minimum clearance required is 0.030 inch and the crank had no trouble clearing. But that wasn't so when a piston and rod assembly was trial-fitted. The block grinding was held to a minimum because of the special-billet Crower stroker rods that use a smaller-length rod bolt, and special clearancing on the bottom of the cap. The bottom of each bore and the pan rail had to be notched, but just slightly. Rich also checked to be sure that the skirt on the piston would clear the

When the boring is done, the block is set up on the Sunnen align-honing machine. By honing (instead of boring) the last couple of thousandths, complete accuracy of main bore size is ensured.

Because we wanted a 4.125-inch diameter, and not the 4.00-inch bore that comes with the Bow Tie block, it had to be custom bored. It took two passes at 0.050 inch and one pass at 0.020 inch on the machine to come close to the final size on the siamese bore block.

Small-Block Outer Limits

crank counterweight in each cylinder.

Next on the agenda were the machining and installation of the four-bolt main caps in the #2, #3, and #4 positions. We used the factory Chevrolet billet caps (PN 14011072) and ARP studs to hold them in place. The photos show how complex the procedure is when it is done right. A special fixture is used to drill the 12-degree angled outer bolt holes after the block has been cut and each cap is hand-fitted to its appropriate main saddle. Hand-fitting ensures that each cap has a tight fit or register in its saddle on the block.

Because each cap is undersized to the main saddle I.D., each main cap must be bored out after it is torqued in place. Shaver's uses oil as a thread lubricant on the ARP studs and bolts, even when installing them in the block for good. It takes three passes on the horizontal boring bar for each main cap and after each pass the cutting tool must be adjusted for the next cut. Very time-consuming!

After the main journals are bored, they need to be honed to their final size. This requires that all caps are torqued into place just as they would be in a running engine. This machining procedure requires a light touch with the honing

Using the 4.00-inch stroke Crower forged crank with the Crower short-bolt billet stroker connecting rods required that only a slight amount of material be removed from the pan rail and the bottom of the cylinder bores. These rods are tops for any type of stroker engine where rod clearance might be a problem.

In order to leave the piston 0.010 inch down in the bore deck clearance, the block deck surface had to be cut 0.030 inch. After the deck is cut, Rich chamfers all the bolt holes to ensure that no threads will pull out.

bar and several checks with the dial-bore gauge. All holes are within 0.0005 inch of being the same size. This is precision machine work at its finest.

Checking the deck surface-to-piston top clearance revealed that 0.030 inch of material had to be removed in order for us to achieve the 0.010 inch down in the bore deck figure we wanted. Using a level and a special gauge that Shaver's made to check deck clearance, Rich fly-cut the deck. By using these procedures, Rich can usually keep the front-to-rear deck taper to within 0.001 inch. When the cutting is done, all the bolt holes on the deck are then chamfered so that no sharp edges will cut up the head gasket.

In order to use the old-style two-piece rear crankshaft seal, a special adapter kit from Chevrolet must be installed. The kit number is PN 10051118 and consists of the adapter and bolts. According to Ron, the key to getting a leak-free seal is to make sure the adapter is aligned properly. To that end, Shaver made a special alignment dowel that locates off number 4 and 5 main caps for a true fit. When the cap has been located properly, Shaver's then drills the adapter and the block at the same time and inserts rollpins to keep the adapter in alignment when the upper half is removed.

The block is then thoroughly washed and dried, after which the cam bearing bores are checked for straightness. This was a problem in some of the first-design Bow Tie blocks, but Ron and Rich assured us that the problem was non-existent in our block. As the final step, the cam bearings are installed in the block. And that is how a cast-iron Bow Tie block is prepped for the outer limits by Shaver Specialties.

Removing the final 0.005-inch of material is accomplished on the Sunnen CK-10 honing machine. Torque plates are bolted to the deck surface of the block and tightened to 50 ft-lbs to simulate the distortion the block goes through when a head is bolted on. This ensures the bores are round and will provide a good ring seal.

The last major machine operation is the installation of the rear main seal adapter. Shaver's uses this special large-diameter dowel to locate the adapter. After this is done, the adapter and the block are drilled and reamed for locating dowels (see text for details).

OUTER LIMITS SMALL BLOCK

Exploring the Short-Block Assembly On A 427-Cubic-Inch Small-Block   Part II

By Jim Losee

aking power is the name of the game in the engine-building business, but power can have two different meanings, depending on whom you're talking to. To most people, power means how much horsepower an engine develops, but to engine builders, the real meaning of power is *torque*. Torque, not horsepower, is what moves a vehicle down the road.

In order to build a substantial amount of torque, the engine must have a large displacement. To that end, we're building a 427cid small-block Chevy. It has plenty of cubic inches to generate torque, and yet it's in a smaller package, both in terms of weight and dimensions, than its big-block counterpart of the same displacement.

If you were with us last month, you read about all the machine work we did on the new-style Chevy Bow Tie small-block with the crew at Shaver Specialties in Torrance, California. This included deburring the block and installing ⅛-inch pipe thread oil galley plugs, machining the main caps and block for 12-degree angled outer bolt four-bolt caps, boring, power-honing, and clearancing the block for the 4-inch-stroke forged Crower crank and special short-bolt stroker rods. The last machining operation Shavers performed was for the installation of the factory rear crank seal adapter that allows the use of a two-piece rear main seal on a block originally equipped with the '87-to-present one-piece seal.

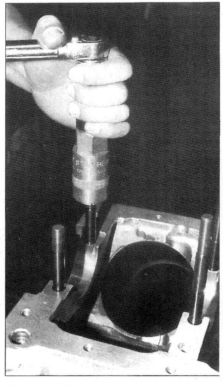

After thoroughly washing the block and drying it, Robert installed the ARP main studs. Using oil as a lubricant and a special stud installation tool, the ARP studs were torqued down to 35 ft-lbs in all positions.

With the main bearing inserts installed in their saddles and caps, the caps were torqued to 45 ft-lbs for a clearance check. Using a dial bore gauge for the inside number, and a digital micrometer for the external measurement, Robert got a 0.003-inch clearance figure. Just right!

In order to get the proper balance in the crank, Crower had to use mallory metal in the counter weights. To aid oiling of the main bearings, the main journal oiling holes on the crank have been chamfered. Also note the very generous radius of the journals, both on the rod and main surfaces of the crank.

Using ARP studs in place of bolts to hold down the main caps created an interference with the Hamburger oil pump sitting flush on the rear main cap. A slight amount of grinding on the pump body gave us the necessary clearance.

To help lubricate the crank under severe operating conditions, G&L Coatings applied their dry-film lubricant coating to the Crower crank. Before the crank is laid in the block, the outer layer of the coating is polished off using a fine Scotchbrite pad.

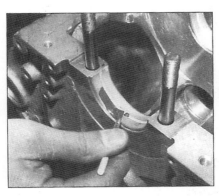

Before the 4-inch-stroke Crower crank is laid into the mains, the bearings are coated with Redline synthetic assembly lube. This helps prevent scuffed bearings and allows a slippery surface for the crank to ride on during initial fire-up.

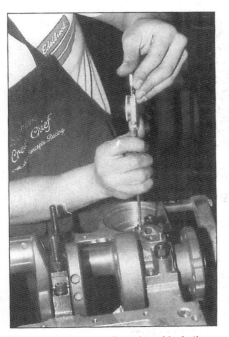

When the factory supplies a bare block, they mean a bare block. The Bow Tie block doesn't come with any of the standard dowel pins, including the ones that locate the oil pump on the rear main cap. These must be 0.500 inch high to locate the pump correctly.

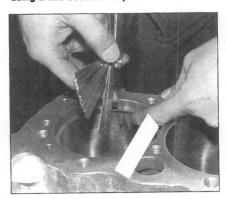

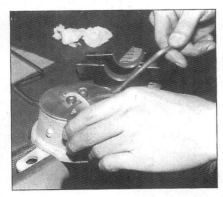

Next, Robert checked ring end gap on the Speed-Pro rings supplied in the Crower stroker kit. The end gap was only 0.006 inch, so they had to be cut to get the 0.014 inch we wanted to run. Rings can be cut by using a ring cutter like the one shown here or they can be filed by hand using a special file to the proper clearance.

With all the machine work done and the block prepped, it's time to assemble the short-block on the "outer limits" 427. After talking with the guys at Alcohol Concepts and our engine builder, Robert Jung, we decided to go with a complete Crower Performance Products stroker kit. Crower's complete rotating assembly kit includes a forged 4340 steel crankshaft, billet stroker connecting rods, custom Ross pistons with Speed-Pro rings and Childs & Albert lightweight wristpins, and specially chamfered main and rod bearings from Vandervell. The complete assembly is neutral balanced from Crower, but to be sure our heavy-duty factory Chevy harmonic damper and Centerforce 30-pound steel flywheel would be compatible with the Crower pieces, we had the whole assembly rebalanced by Automotive Balancing Service of South Gate, California.

Adequately oiling an engine of this

Holding the high-volume Hamburger oil pump in place is an ARP oil pump stud and 12-point nut and flat washer. Using a stud here won't distort the rear main cap or cause a wiped-out bearing.

Small-Block Outer Limits

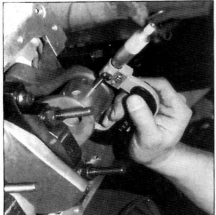

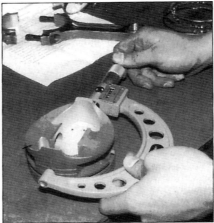

Because of the long 4-inch stroke and the 0.150-inch longer-than-stock connecting rod, the wristpin hole in the piston must be drilled into the oil control ring land. To support the ring in the pin area, a special support ring must be used.

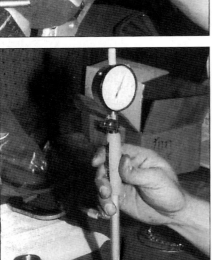

Before installing the rings on each piston, it is wise to check whether the ring will fit snugly in its groove. Robert checks each one and then installs the rings, either by using a ring spreader, or by *carefully* wrapping the ring into its appropriate groove.

For a high-performance engine to live and be reliable, all clearances need to be checked and double checked. Here, Robert is checking rod bearing clearance. All rods had 0.0025 inch clearance, plus or minus 0.0002 inch. That's two ten-thousandths of an inch! Pretty precise, right?

Next on the list to be checked was the piston-to-cylinder bore clearance. Although Shaver Specialties bored each cylinder to a specific piston size, Robert always double checks. Each cylinder had 0.006 inch clearance. Perfect!

Robert prefers to use a B&B tapered piston ring compressor for installing the piston/rod combo into the cylinder. It's smooth and easy. The dull finish on the pistons is G&L Coatings' dry-film lubricant coating which helps to reduce friction between the piston and bore. This is especially helpful when the piston is at bottom dead center and is pushing on the side of the cylinder as it is trying to go up.

size is very critical. Having the oil at the wet-sump pump pickup is a result of how well the pump works and the capacity of the oil pan itself. To this end, we enlisted the aid of one of the most experienced producers of oil pans and pumps anywhere, Ed Hamburger. After discussing the parameters of the "outer limits" 427, Ed sent us one of his all-aluminum stroker, low-ground-clearance, internally baffled and windage-tray-equipped, 7-quart wet-sump pans. Along with the pan came one of his high-volume, high-pressure oil pumps and pickup. The really trick feature of the pan is its special cast aluminum rail that is clearanced for stroker cranks, and yet has plenty of strength to seal the pan to the block without flexing and causing a leak. The pan is sealed to the block using Fel-Pro gas-

In order to hold the full-floating lightweight Childs & Albert wristpins in the piston, two Spiralocs are used on each end of the pin. This ensures that the pins won't come out under severe loading. They are a pain to put in, but the same could be said if the pins came out.

kets and ARP pan bolts.

Timing the Crower steel billet roller camshaft with the crank is a Summers Brothers gear drive. The gear drive provides very accurate timing and allows us an easy way to advance or retard the cam for maximum power. The only

Small-Block Outer Limits

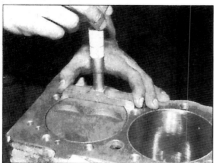

Even though we knew that the block had been decked for 0.010-inch clearance by Shaver Specialties using their special depth micrometer, Robert double checked anyway. Of course, it was right on.

mods that have to be done when using this gear drive is that all the pulleys and mounting brackets must be spaced outboard approximately 3/8-inch and the oil pan seal lip on the factory steel timing cover must be removed in order to seal tightly to the Summers gear drive adapter plate.

Now it's time to follow along with us as Robert from Alcohol Concepts assembles our "outer limits" 427 short-block. One of the key lessons to be gleaned from this assembly is that you should always double-check every dimension. Remember the old adage: it's better to be safe than sorry.

To aid header clearance in our Camaro's chassis where the 427 will eventually rest, we used a Traco oil filter adapter provided by Hamburger. It will allow us to put the filter out of the way and yet still be accessible.

When inserting a steel roller cam in the block, Alcohol Concepts coats each cam bearing with Redline synthetic assembly lube, and then they coat each cam lobe with regular motor oil.

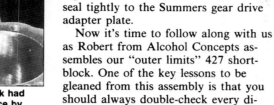

When installing the cam gear for the Summers Brothers gear drive, assemble the Torrington needle-thrust bearing and washer in the correct sequence. If there are any questions, refer back to the instruction sheet, which of course should have been read before attempting installation.

These little Woodruff keys prevent the crank timing gear and damper from rotating freely. They should be installed in the keyways on the crank using a brass-headed hammer only. A steel-headed hammer will flatten the Woodruff keys and won't allow the gear or damper to be installed.

With the right amount of clearance between the crank gear and the crank snout recommended by the instructions, the Summers Brothers gear drive crank gear can be installed using a brass drift and a hammer. Tap in the gear until it sits flush with the front of the crank.

Just in case there is an unlikely valvetrain failure, Robert installed the Hamburger oil return hole screen kit to prevent any unwanted shrapnel from finding its way down to the oil pan and pump.

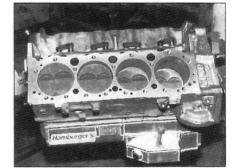

Here's the short-block assembled and ready for the final build. Note the aluminum 7-quart Hamburger oil pan which features a thick cast aluminum pan rail for a leak-free fit using Fel-Pro's latest pan gasket set.

SOURCES

Automotive Balancing Service
Dept. CC
P.O. Box 1984
South Gate, CA 90280
213/564-6846

ARP/Automotive Racing Products
Dept. CC
250 Quail Court
Santa Paula, CA 93060
800/826-3045
In CA, 805/525-5152

Crower Performance Products
Dept. CC
3333 Main Street
Chula Vista, CA 92011
619/422-1191

Fel-Pro Inc.
Dept. CC
7450 No. McCormick Blvd.
Skokie, IL 60076
312/761-4500

G&L Coatings
Dept. CC
888 Rancheros Dr., Unit F2
San Marcos, CA 92069
619/743-0224

Hamburger's Oil Pans
Dept. CC
1501 Industrial Way North
Toms River, NJ 08755
201/240-3888

Midway Industries Centerforce Clutch
Dept. CC
P.O. Box 980
Midway City, CA 92655-0980
714/898-4477

Summers Brothers Inc.
Dept. CC
530 So. Mountain Ave.
Ontario, CA 91762
714/986-2041

OUTER LIMITS SMALL BLOCK

The Final Assembly Of The Mega-Cube Small-Block Chevy   Part III

By Jim Losee

We're down to the short strokes now on the 427cid small-block. The guys at Alcohol Concepts will be installing the ARP head studs and topping off the short-block with the Brodix Track I heads, the Edelbrock Victor Jr. intake manifold using Fel-Pro gaskets, and the MSD distributor.

Robert Jung of Alcohol Concepts says that there are several very important things to do when finishing off an engine like this, and, as we have discussed in past installments, one of the most important is to double-check everything for both proper fit and clearance.

First, we installed an Edelbrock polished cast-aluminum front cover for the Summers Brothers geardrive. Being cast aluminum, it will help reduce some of the noise associated with geardrives and it provides a point stiff enough for the roller cam button to contact if the cam starts to walk. In order for this cover to fit on the Summers geardrive, we had to mill the lip that would normally ac-

cept the front oil pan seal, as that lip is now on the Summers geardrive support. Not only does the Edelbrock front cover have several functional purposes, it also looks good.

Another Edelbrock product we used while building the 427 short-block was their freeze plug, oil galley plug, and head dowel kit. This kit comes complete with brass freeze plugs, ⅜-inch

To help keep noise down and provide a solid stop for the cam button, we used an Edelbrock polished aluminum front cover. In order to fit the Summers geardrive, we machined off the oil pan seal lip and bolted it directly to the geardrive support.

pipe screw-in allen-socket oil plugs, and the head dowels to replace those that are often lost or destroyed when a block is decked. We like this kit because it has all of these plugs and dowels in one package and is available from a single source.

An air pump with 427 cubic inches requires a decent size and flowing set of cylinder heads for maximum efficiency

102

Small-Block Outer Limits

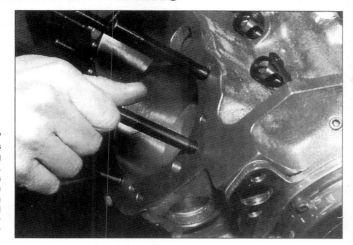

Installing the ARP head studs is as easy as screwing them into the block. ARP recommends that the studs be screwed-in hand-tight, and that's all. Pretty easy, huh?

Because our block had to be decked and it didn't come with head dowel pins from the factory, the plug-and-dowel kit from Edelbrock came in handy. Robert Jung of Alcohol Concepts installs the dowels with a hammer and leaves them sticking ¼-inch out of the deck surface.

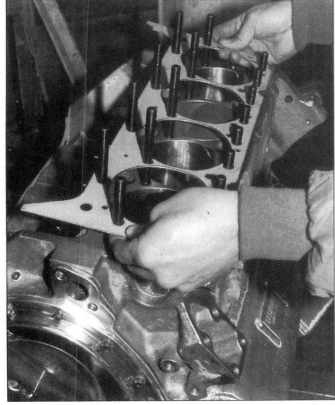

A pair of Fel-Pro Perma-Torque blue head gaskets seal the combustion chambers to the block. Robert applies a dab of Gasgacinch around each of the water holes on the deck before dropping the gasket in place.

and power. To that end, Alcohol Concepts suggested we use a pair of the latest offering from Brodix, the Track I. With its 215cc intake port, 76cc combustion chamber, and standard 2.05-inch-diameter intake and 1.60-inch-diameter exhaust Manley Pro Flow valves, this head should allow us to make almost 550 hp. Another trick thing about the Brodix Track I head is that it has a raised exhaust outlet that exits from a "D"-shaped exhaust port, helping to remove exhaust gases quicker.

We got our heads bare with only the seats and heli-coils installed in the bolt holes and had Port Flow Design of Carson, California, assemble these heads with the Manley valves and ARP rocker studs. The stock Chevy guideplates were installed after doing one of their bowl and gasket match porting jobs. Rick Kemph of Port Flow Design says these are some of the best-flowing out-of-the-box aluminum heads he has ever seen. You can also get the Track I heads completely assembled with springs and valves with a mild porting

If all the studs are in their proper place and they're straight, the Brodix Track I heads should just drop down over the head studs.

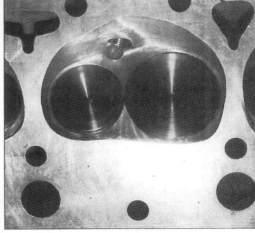

The aluminum Brodix Track I head comes standard with a 76cc combustion chamber and 2.05-and 1.60-inch-diameter intake and exhaust valve seats, respectively. To gain the most in fuel/air flow into the chamber we used a set of Manley Pro Flow stainless valves with undercut stems.

Small-Block Outer Limits

To optimize the capabilities of the Brodix Track I heads, Port Flow Design in Carson, California, did some mild bowl and porting work to the already good-flowing heads. The standard intake port from Brodix features a 215cc intake runner. The exhaust has a raised "D"-port shape for good flow characteristics.

job directly from Brodix.

Inducting air and fuel into the cylinder heads is done via the most versatile intake manifold of all time, the Edelbrock Victor Jr. This manifold does so many things so well on such a variety of engine combinations that we have come to use it almost exclusively, especially on big-cube small-block Chevys. With just a port match and a 1-inch carb spacer, this manifold yields the most torque and horsepower for a street or mild race engine of any around.

Doing the fuel and air mixing on our 427 is a Holley 8804 830cfm carb. The folks at Fuel Curve Engineering recommended this carb over the normal 4781 850cfm unit because it has 1-11/16-inch-diameter throttle bores versus the 1-3/4-inch-diameter bores on the 4781. The smaller bore will provide better response down low than the bigger carb. Fuel Curve also made their Saturday Night Special modifications to our carb to increase the power and airflow capabilities of the engine.

In order for the rods to clear our

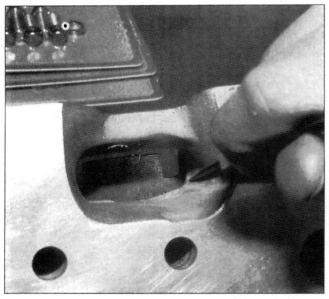

Before the guideplates for the pushrods have been bolted down, check the clearance between the intake runner wall and the pushrod. Sometimes all you have to do is move the guideplate a little one way or the other for the added clearance; in some instances, a slight amount of aluminum can be ground off for clearance.

Before inserting the pushrods in the clean engine, wash them off in a solvent tank and blow them dry with compressed air. This will eliminate the chance that anything from inside the pushrod tube will get into the oil flow in the block.

Holding the rocker arms in place is a set of ARP 7/16-inch diameter screw-in studs. Providing pressure for the opening and closing of the valves from the cam is a set of Crower roller cam valvesprings that have 200 psi of seat pressure closed and 475 psi open.

When the heads are torqued down using studs, make sure you use a criss-cross bolt-tightening pattern and bring them down to 45 ft-lbs in 15-psi increments.

Here, Robert Jung from Alcohol Concepts is adjusting the clearance between the valve and the Crower stainless steel roller rocker arm. In order to do this correctly, it practically needs three hands, but if you're careful, it can be done with two hands as shown.

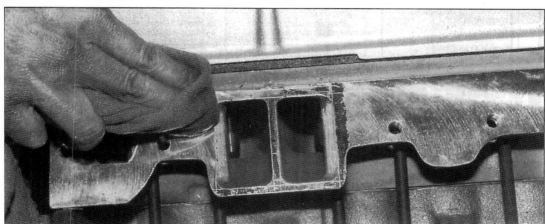

Make sure the gasket surface around the intake port on the head is clean so the gasket will adhere correctly.

The guys at Alcohol Concepts don't believe in using end seal gaskets. Instead, they use a large, smooth bead of RTV from Fel-Pro and let it set up about 10 minutes before putting the intake manifold in place.

The Port Flow Design port-matched Edelbrock Victor Jr. intake manifold is laid down with care on the heads against the #1206 Fel-Pro Print-O-Seal intake gasket. The best way to install the manifold is to lay it straight down so that it touches both heads at the same time for a better seal.

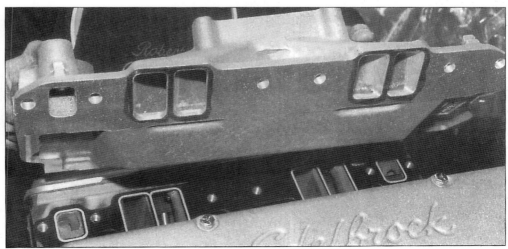

Small-Block Outer Limits

Crower cam (duration at 0.050-inch; 254 degrees intake and 260 degrees exhaust with a valve lift of 0.621-inch intake and 0.624-inch exhaust cut on 108 degree lobe centers), we installed a roller cam that had a 0.925-inch-diameter base circle. The normal roller cam comes with 1.000-inch base circle and the rods (even the Crower stroker rods with a shorter bolt) will still hit the cam with a 4.00-inch stroke crank and prevent the engine from rotating. Thus the smaller-than-normal base circle cam must be used. This slightly reduces the cam's strength, but when you're only twisting it to 6000 rpm, it will work fine.

Another problem that crops up when using a smaller base circle cam is that the valvetrain geometry goes away and the rocker arms end up hitting the edge of the valvespring retainers when the lifter is traveling to the heel, or bottom, of the cam. This can create severe valvetrain problems such as pushrods coming out of their seats on the rocker and bending.

In order to solve this type of problem, Crower supplied us with an adjustable pushrod so that we could have the proper-length pushrods made for the smaller base circle cam. You adjust the pushrod until all the geometry comes back into sync throughout the full rotation of the camshaft, and then measure the length and have a full set of pushrods made to that spec. In our case, it took a pushrod that was 0.500-inch longer than stock to put the geometry back on the right plane. For longevity's sake, we also had the pushrods coated by G&L Coating with their dry-film lubricant to reduce wear on the pushrod as it rubs on the guideplate.

In a couple of months we'll have baseline power numbers with the carburetor and then we're going to install the Street Legal Performance fuel injection system on this monster and stretch the bounds of power and torque for a small-block Chevy on the street.

An important point to consider here is that there are two "O"-rings that go on the bottom of the MSD distributor. These "O"-rings seal the oil galley in the block, ensuring that all the oil goes to the main bearings. Read the directions on this distributor before you wash the block and start building the short-block.

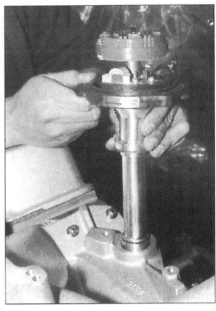

Before setting the MSD distributor down in the block, be sure that a bronze distributor gear is used against the steel roller cam and be sure the bronze gear is one from MSD that has a 0.500-inch I.D. hole. A standard 0.450-inch I.D. bronze gear won't work.

SOURCES

Automotive Balancing Service
Dept. CC
P.O. Box 1984
Southgate, CA 90280
213/564-6846

Automotive Racing Products
Dept. CC
250 Quail Ct.
Santa Paula, CA 93060
805/525-5152

Brodix Cylinder Heads
Dept. CC
P.O. Box 1347
Mena, AR 71953
501/394-1075

Centerforce Clutches/Midway Ind.
Dept. CC
P.O. Box 980
Midway City, CA 92655-0984
714/898-4477

Chevrolet Performance Parts
Available at over 4500 Chevrolet dealers nationwide

Crower Racing Products
Dept. CC
3333 Main St.
Chula Vista, CA 92011
619/422-1191

Edelbrock Corp.
Dept. CC
2700 California St.
Torrance, CA 90503
213/781-2222

Fel-Pro Inc.
Dept. CC
7450 N. McCormick Blvd.
Skokie, IL 60076
708/674-7700

Fuel Curve Engineering
Dept. CC
Route #2, Box 35
Tryon, NC 28782
704/894-3511

G&L Coatings
Dept. CC
888 Rancheros Dr.,
Unit F2
San Marcos, CA 92069
619/743-0224

Hamburger's Oil Pans
Dept. CC
1501 Industrial Way N.
Toms River, NJ 08755
201/240-3888

Holley RPD
Dept. CC
11955 East Nine Mile Rd.
Warren, MI 48089-2003
313/497-4250

Manley Performance Products
Dept. CC
1960 Swarthmore Ave.
Lakewood, NJ 08701
201/743-6577

Port Flow Design
Dept. CC
348 C East Carson St.
Carson, CA 90745
213/835-4457

Shaver Specialties
Dept. CC
20608 Earl St.
Torrance, CA 90503
213/370-6941

Summers Brothers, Inc.
Dept. CC
530 South Mountain Ave.
Ontario, CA 91762
714/986-2041

Here's the final product: 427 inches of rompin', stompin' small-block Chevy! We'll give you dyno results in about two months.

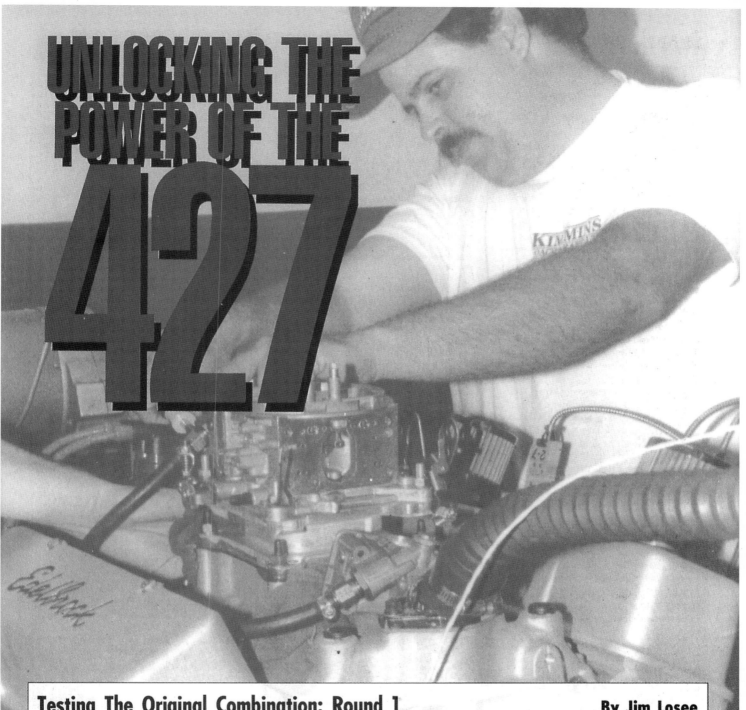

Unlocking the Power of the 427

Testing The Original Combination: Round 1

By Jim Losee

It's one of those questions that is going to get asked eventually and until right now we haven't had a definitive answer. The question: How much streetable power will the 427 make? We've been quizzed about this quite a bit and have been as unsure about the issue as you. Well, now we're going to find out.

As you saw in the June '90 issue of CC, we expressly constructed a 427cid small-block Chevy to test the performance limits of a carbureted large-displacement street small-block. This round of testing is devoted to the results of the carbureted test session; the results of the cam change, carb spacer, and exhaust crossover pipe tests will appear next month.

The parameters of this test required that the engine combination would have to be driveable for street use. To that end, when we were selecting a carb for the 427 engine combination, we consulted with the folks at Fuel Curve Engineering in Tryon, North Carolina, about what type and size of carburetor we should run for this test. Bob Szabo suggested that we use a Holley PN 8804 830cfm carb with dual accelerator pumps and large boosters. To retain the driveability factor and yet get all the performance we could from the Holley carb, Fuel Curve massaged the 8804 with their "Saturday Night Special" mods that boosted the total airflow of the carb from 830 cfm to just a hair over 900. Also included in these mods is the addition of four-corner idle circuits that help an engine with a rather large camshaft, such as the Crower piece we're running, to idle better.

As you saw in the conclusion of the 427 buildup, we chose to run the Edelbrock Victor Jr. single-plane intake

Unlocking The Power Of The 427

manifold that had been port-matched to the Fel-Pro PN1207 intake gaskets by Port Flow Design in Carson, California. This is one of the most versatile manifolds to come up the pike in a long time. Because of the size of our test engine, we were right at the edge of the performance envelope for the Victor Jr. To gain every little bit from the manifold, the crew at Alcohol Concepts blended the plenum area and radiused the entryways to the runners.

Not knowing what to expect in the way of power from the 427, we decided to run the carb/manifold combination the first couple of times without a 2-inch open carb spacer. Although we thought that using the spacer would probably be the right combination for the most power, which it was, we wanted to test every option available to us.

Some of the other avenues we explored were changing valve lash, both loosening and tightening it beyond the Crower specs, changing to a full-on Winston Cup/NASCAR 750cfm carb (which showed negligible gains), changing the timing from 32 degrees total to 40 and back down to 30 in two-degree

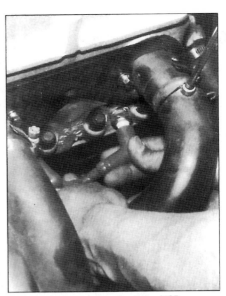

With the generous length of the MSD shielded and sealed plug wires, we were able to run the wires underneath the exhaust headers. The holes in the tops of the header tubes are for the exhaust gas temperature probes that are held in place with stainless wire wrapped around the header tube.

In order to get the oil in and out of the engine with the least amount of restriction, we used -12 AN stainless overbraided lines coming out of the Traco/Hamburger adapter. The Hamburger aluminum pan held 7 quarts of oil, which was measured on the pan dipstick, because Bow Tie blocks don't come with either right or left dipstick tube holes drilled.

To gain the most power from the Victor Jr. intake manifold, the crew at Alcohol Concepts blended and radiused the plenum and entries to the runners. The carb opening was also opened up to gain maximum velocity from the carb mixture.

Sitting atop our Edelbrock Victor Jr. intake manifold is the latest in water outlet gaskets from Fel-Pro. This is a really trick piece as it has a hard plastic spacer that has two receiver grooves (one on top and one on the bottom), which hold a wide O-ring for sealing. It's also reusable.

Here's the 427 just after it was settled onto the dyno stand. Notice the deep-groove factory Chevrolet steel pulleys and the heavy-duty GM belts that connect the crank and water pump. All are available over-the-counter at your Chevrolet dealer. Port Flow Design welded the aluminum breather tubes to the Edelbrock valve covers for us, NASCAR-style.

TEST 1

CC 427 CHEVY SB
0.024I/0.026E VALVE LASH HOT
2975 MANIFOLD 8804 HOLLEY/FCE CARB
BRODIX TRACK1 HEADS 1⅞ HDRS 32T
500-RPM STEP TEST

RPM	CBT lb-ft	CBHP	VE%	A/F	BSFC lb/HpHr
2500	453.6	215.9	89.5	12.1	.48
3000	478.0	273.0	89.4	11.8	.47
3500	483.5	322.2	90.2	11.8	.47
4000	474.0	361.0	102.8	11.7	.48
4500	533.0	457.1	102.4	12.0	.47
5000	528.9	503.5	104.4	12.0	.49
5500	498.1	521.6	103.2	12.1	.51
6000	455.7	520.6	100.4	12.2	.54

Test #1 was our initial baseline test after breaking in the 427. The tests were done on the engine as received from Alcohol Concepts with no changes. Note the timing, valve lash, and the huge jump in power in the mid-range starting at 4500 rpm.

Unlocking The Power Of The 427

increments (with best power showing at 34 degrees), changing spark plug gap (with no power gain evident), and jetting the carburetor several different times to achieve the most power.

As you can see, hardly any stone was left unturned in the quest for the ultimate in performance with our current

Robert Jung of Edelbrock is seen tightening the fuel line nuts on the carb bowls. Use caution when doing this as the bowls will strip easily when its fuel line nuts are overtightened.

TEST 2
CC 427 CHEVY SB
0.024I/0.020E VALVE LASH HOT
2975 2-INCH OPEN SPACER
8804 HOLLEY/FCE CARB W/78/84 JETS
BRODIX TRACK 1 HEADS 1⅞ HDRS 34T
500-RPM STEP TEST

RPM	CBT lb-ft	CBHP	VE%	A/F	BSBFC lb/HpHr
3000	472.6	270.0	88.9	12.1	.46
3500	479.4	319.5	90.1	12.2	.46
4000	464.9	354.1	89.7	11.8	.48
4500	537.8	460.8	103.7	12.4	.46
5000	536.4	510.7	106.3	12.3	.48
5500	505.7	529.6	105.2	12.7	.49
6000	469.1	535.9	101.6	12.6	.51

Test #2 illustrates the best power we could get from the 427 utilizing all the changes and tuning tricks we could think of. Again, note the timing and valve lash settings. And as in the previous test, notice the large jump in torque and horsepower between 4000 and 4500 rpm. We still aren't sure what caused this, but we're working on it and will let you know.

In an effort to find more power we even tried one of Edelbrock's full-on Winston Cup/NASCAR Holley test carbs. It is rated at 750 cfm, but if you look very closely at it, you'd swear it was a bunch bigger. This carb didn't show any gain.

Fuel Curve Engineering prepared our 8804 830cfm Holley carb by doing their "Saturday Night Special" treatment. This brought the cfm rating to a little over 900 and it added a four-corner idle adjustment for use with a large cam like our Crower part. It helped the idling characteristics greatly.

Unlocking The Power Of The 427

cam/carb/manifold combination. We feel that we've set a benchmark for comparison. With all the testing we've done, we believe that the 427 has made the best power possible with the combination that's being used.

But what about the almost nitrous-like boost in horsepower between 4000 and 4500 rpm? We can't really explain what's going on in that rpm range, but boy, does the 427 sound like a whole 'nother engine from 4500 to 6000. (Next month in Part II of our testing, we will attempt to solve this dilemma.)

With help from the great folks at Edelbrock who allowed us the generous use of their dyno facilities and their time, especially Robert Jung, Mike Eddy, Jack Ring, and Curt Hooker, we'll figure the problem out.

For those of you who are wondering why we stopped testing at 6000 rpm, it's not for a lack of trying. We ran a couple of tests past our self-imposed 6000 limit to 6500 just to see what would happen and came away with the conclusion that there wasn't any power to be gained up there.

Another test result we included that we haven't in the past is the 600-rpm-per-second acceleration run. This test more closely approximates how the engine will react in the car and what kind of power it's making as it is pulling through each gear. As you can tell by looking at the results of the acceleration test (test #3), the 427 makes ample power (both torque and horsepower), to overcome almost any type of street tire available, and probably a goodly portion of the slicks that will fit within the confines of a stock wheel opening.

Follow along with us as we show you what was involved in doing the carb/manifold test, as well as the results. And stayed tuned for the September issue, as we search for even more streetable power in the 427.

There are two things to note here. (1) The 2-inch open-plenum Edelbrock carb spacer, which helped us gain the most torque and horsepower from the 427, and (2) the use of the MSD 6A ignition system, including the distributor. In our power quest we also played with timing, going up and down from our initial setting of 32 degrees total and found that 34 was where the 427 liked it best.

TEST 3
CC 427 CHEVY SB
0.0241/0.020E VALVE LASH HOT
2975 MANIFOLD 2-INCH OPEN SPACER
8804 HOLLEY/FCE CARB W/78/84 JETS
BRODIX TRACK 1 HEADS 1⅞ HDRS 34T
600-RPM/SECOND ACCELERATION TEST

RPM	CBT lb-ft	CBHP	VE%	A/F	BSBFC lb/HpHr
2500	450.0	214.2	90.3	12.6	.47
2750	426.1	223.1	83.9	12.2	.47
3000	441.9	252.4	82.2	13.0	.42
3250	470.0	290.8	85.6	13.2	.40
3500	468.7	312.3	87.0	12.9	.42
3750	452.3	322.9	85.5	12.9	.43
4000	445.3	339.1	85.1	12.7	.44
4250	465.5	376.7	87.4	13.8	.40
4500	522.5	447.7	98.4	14.5	.38
4750	523.8	473.7	102.0	14.3	.40
5000	527.8	502.5	104.9	13.8	.42
5250	513.0	512.8	104.2	13.6	.44
3500	490.6	513.8	102.1	13.4	.45
5750	478.8	524.2	101.5	13.6	.46
6000	456.5	521.5	100.4	13.3	.49

Test #3, the 600-rpm-per-second acceleration run, most closely approximates what power levels the engine will have in a car when you jump on the throttle. The "outer limits 427" will definitely smoke the tires from the start, but when it hits 4500 rpm it'll feel like somebody kicked in a 100-horse nitrous unit. And does it make some torque, even at 6000 rpm!

To get every last ounce of power out of the 427, we adjusted the valves both looser and tighter than the original Crower specs to see if minutely changing the duration would help. It didn't help, so we reset the stainless Crower rockers at the original lash setting of 0.024-inch intake and 0.026-inch exhaust.

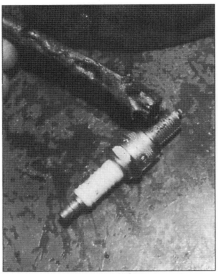

We even tried changing the plug gap for some additional power but didn't see any. A critical thing to remember when putting spark plugs in aluminum heads is to use some sort of anti-seize compound on the threads of the plugs so they won't weld themselves to the aluminum. The plugs are steel and the threads in the heads are aluminum.

Mike Eddy of Edelbrock is seen here helping to change jets one more time on the 8804 Holley carb. After several changes, the most power was seen using #78 jets in the primaries and #84's in the secondaries.

UNLOCKING THE POWER OF THE 427

Round 2

Reaching The Outer Limits Of Performance

By Jim Losee

Last month, we described the first round of extracting the most power we could from our "Outer Limits" 427 small-block Chevy. We tried a couple of different carburetors, different timing, a multitude of jet changes, different lash settings, and a 2-inch open-plenum carburetor spacer. The results were pretty good, but not quite as much as we had in mind. And besides, there was that tremendous jump in horsepower and torque from 4000 to 4500 rpm to contend with.

After talking things over with the crew at Edelbrock, cam grinder Dave Crower, and Fuel Curve Engineering's Bob Szabo, we decided that we should try another cam to see if we could get rid of that 100+ horsepower leap at 4500 rpm. Along with the cam change, we wanted to try a different combination of spacers, which didn't show much difference, and install an exhaust crossover pipe between the collectors. This last item did change power levels, but it also showed us where a major deficiency lay.

Dave sent us another steel billet roller cam that was similar in specs to the cam we were currently using. The thinking was that we would take away some power at the top end and increase it at the bottom by going with a wider lobe spacing and a bit shorter duration and lift.

When we took the 427's upper end apart to replace the cam, the engine looked like it hadn't even been run. Everything was clean as a whistle and there wasn't a trace of metal in the filter or the valley screens. The Crower roller lifters and pushrods looked great, with no signs of wear, and the intake ports in the heads were as dry as a bone.

After checking everything else and not finding any problems, the Edelbrock gang installed the new roller cam. The cam was inserted with 4 degrees of advance using a Summers offset bushing. Buttoning up the rest of the engine was a snap because it was on the dyno. I wish cam changes in a car could be done half this easily.

After the initial warm-up of the 427 with the new cam, and a subsequent hot valve lash adjustment, Robert Jung and Matt Compton made a steady-state power run at 4000 rpm to make sure everything was compatible. Everything looked fine, but we still had a low power reading at the same rpm that we had with the previous cam.

On to the step tests. The first one consisted of taking two 1-inch spacers, one with four holes in it and one with an open plenum, and stacking them on

Unlocking The Power Of The 427

When we took the top end of the 427 apart for our camshaft change, everything in the upper end looked super. There were no metallic particles anywhere, not even when the oil filter screen was cleaned. The Crower roller lifters looked like they hadn't been used.

top of each other, and making a run. The reason this was tried was that using the four-holer on top and the open plenum on the bottom would isolate any reversion signal at the base of the carburetor and give the air/fuel mixture a better direction toward the intake manifold runners. This combo was a bit off from our 2-inch open spacer, so we tried an assortment of 1- and 2-inch spacers, all with relatively lackluster results. The combination the 427 liked was the 2-inch open spacer and that was it.

During all this testing we noticed that the new cam that we thought would increase the bottom-end power and decrease the top-end power did exactly the opposite. Power at the top end was increased and the power at the low end went down. We were glad to see the upper-end power pick-up, but con-

CC 427 CHEVY SB   TEST 1A
2975 MANIFOLD 8804 HOLLEY/FCE CARB
NEW CAM, TWO 1-INCH SPACERS
500-RPM STEP TEST

RPM	CBT lb-ft	CBHP	VE%	BSFC lb/HpHr
2500	446.2	212.4	86.8	.47
3000	467.9	267.3	87.4	.44
3500	479.0	319.2	90.1	.45
4000	479.6	365.3	90.0	.44
4500	537.1*	460.2	103.5	.44
5000	530.4	505.0	106.2	.46
5500	510.0	534.1	104.8	.47
6000	478.8	547.0*	101.5	.49

* Denotes power peaks in each test

CC 427 CHEVY SB   TEST 1B
2975 MANIFOLD 8804 HOLLEY/FCE CARB
NEW CAM, TWO 1-INCH SPACERS
300-RPM/SEC ACCELERATION TEST

RPM	CBT lb-ft	CBHP	VE%	BSFC lb/HpHr
2500	437.1	208.1	86.6	.47
2750	433.9	227.2	83.4	.44
3000	449.7	256.9	83.3	.41
3250	479.7	296.8	87.2	.40
3500	477.0	317.9	88.1	.43
3750	458.2	327.2	86.9	.44
4000	463.0	352.6	87.5	.44
4250	492.3	398.4	91.4	.39
4500	529.1	453.3	100.6	.42
4750	535.8*	484.6	103.9	.42
5000	529.2	503.8	104.0	.43
5250	511.9	511.7	103.8	.44
5500	501.5	525.2	103.3	.46
5750	486.4	532.5*	101.5	.45
6000	465.8	532.1	100.1	.49

* Denotes power peaks in each test

CC 427 CHEVY SB   TEST 2A
2975 MANIFOLD 8804 HOLLEY/FCE CARB
NEW CAM, 2-INCH OPEN SPACER
3.5-INCH EXHAUST CROSSOVER
500-RPM STEP TEST

RPM	CBT lb-ft	CBHP	VE%	BSFC lb/HpHr
2500	430.5	204.9	84.1	.46
3000	470.9	269.0	87.9	.44
3500	463.7	309.0	87.0	.44
4000	500.4	381.1	95.2	.46
4500	539.2*	462.0	103.6	.45
5000	535.8	510.1	104.9	.45
5500	512.2	536.4	104.2	.48
6000	474.2	541.7*	101.3	.49

* Denotes power peaks in each test

CC 427 CHEVY SB   TEST 2B
2975 MANIFOLD 8804 HOLLEY/FCE CARB
NEW CAM, 2-INCH OPEN SPACER
3.5-INCH EXHAUST CROSSOVER
300-RPM/SEC ACCELERATION TEST

RPM	CBT lb-ft	CBHP	VE%	BSFC lb/HpHr
2500	420.7	200.1	83.5	.47
2750	418.0	218.9	81.6	.42
3000	451.0	257.6	82.6	.41
3250	472.4	292.3	86.2	.40
3500	455.7	303.7	85.5	.43
3750	446.2	318.6	85.8	.45
4000	490.4	373.5	92.0	.41
4250	513.4	415.5	96.9	.41
4500	528.7*	453.8	101.7	.42
4750	527.0	476.6	102.8	.43
5000	527.1	501.8	103.6	.44
5250	516.8	516.6	103.9	.46
5500	505.6	529.5	103.2	.47
5750	485.2	531.2	102.5	.48
6000	469.4	536.3*	100.6	.50

* Denotes power peaks in each test

When perusing the dyno charts, take a look at where the horsepower high jump takes place. With the crossover, it comes in at a lower rpm. Also note that we used a 300-rpm/sec acceleration test, not a 600-rpm/sec one as we did in Round I. After talking to several engine builders and dyno operators, we felt a 300-rpm/sec test was the most representative of street car acceleration.

The cam was dialed in with a 4-degree advance by using a Summers bushing. The new cam was milder than the previous one and we thought it ought to pump up the bottom and take a bit of power away from the top. Were we wrong! Compare this issue's numbers with CC's August '90 test results.

While the Edelbrock Victor Jr. intake was off for the cam change, the crew massaged it on the floor of the plenum. The mods weren't readily noticeable, but Edelbrock says every little bit helps.

versely we didn't want to lose any low-end power either. We wondered if we couldn't have our cake and eat it, too.

Besides the power pick-up, we could get the 427 to idle down to a rumpity 850 rpm with no load on it. This is as low as we could go and it simulated what would happen in a manual or high-stall-speed-equipped automatic transmission car. And the vacuum was at 10 inches, enough to operate power

Unlocking The Power Of The 427

accessories, including brakes. With the idle set at 1000 rpm, the engine had 11 inches of vacuum and idled with some smoothness. When running a cam with this much duration at 0.050-inch, idling and smoothness are kind of mutually exclusive, and relative only in the eyes of the beholder. Between the whine of the Summers geardrive and the rumpity-rump of the cam, it's that type of idle sound that brings a big smile to the face of a knowing street machiner.

The last test we ran involved installing a 3-inch-diameter crossover pipe after the collectors. After Matt Compton fabricated the crossover and Robert Jung adjusted the valves and came up with the correct length of flexible steel exhaust ducting, we ran the 4000-rpm steady-state test. Lo and behold, we had eliminated the hole at 4000 and had the torque number up to 512 lbs/ft from a previous best of 480. It looked like we had found our cure to the horsepower high jump.

But not so fast. When we did the step test we found that all we had done was lower the horsepower high jump from 4500 rpm to 4000 rpm. Now we're not saying we hadn't done something positive, because we decreased the jump in power from 110 horses to "only" 70. Better, but not good enough.

After all this testing, we finally ran out of time, both on the dyno and for

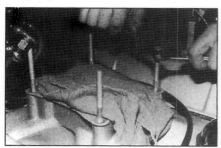
Here's the first spacer combination we tried with the new camshaft: a 1-inch four-holer on top and a 1-inch open on the bottom. We felt that these spacers might decrease the horsepower high jump at 4500 rpm.

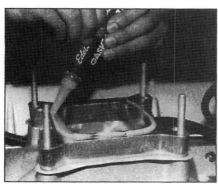
Before installing the two 1-inch spacers, we had to change the carb studs to 3-inch-long ones. If you don't have a stud installer, double-nut it just as we're doing here.

The crew at Edelbrock even went so far as to make sure that the openings of the four-hole spacer were as big or bigger than the venturis on the Fuel Curve Engineering-modified Holley 8804 900cfm carb, before the adapter was installed. Even though it doesn't look like it, there really is a very thin bead of silicone to seal the two spacers.

Jetting on the FCE Holley 8804 was left the same as in the previous test: 78's in the front and 84's out back. The little bracket at the right is to retain the SuperFlow airflow meter. There's also one diagonally opposite.

So that there would be a solid seal between the two 1-inch spacer plates, a small bead of silicone was laid down on the top of the bottom spacer. For best mixture flow, compare the spacer opening with that of the manifold. We cleaned up both our single 2-inch open and 1-inch open spacers for a better match.

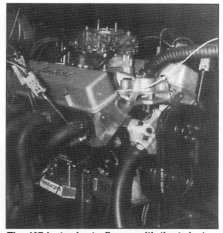
The 427 just prior to fire-up with the twin 1-inch spacers. After any type of carb removal and reinstallation on a dyno engine, the Edelbrock gang checks for closed needles and seats with a quick hit from the electric fuel pump.

Here's the installation of the 2-inch open carb spacer for the second time. This combination proved to be the one the 427 liked the most by producing the most power.

Unlocking The Power Of The 427

story deadline. But we learned that our stumbling block was the exhaust system and that is what caused the horsepower high jump. The headers we used were 1.875 inches in diameter and are as big as one can go utilizing the stock Chevrolet exhaust bolt pattern. This tube size wasn't quite big enough to stop some exhaust pulsing back up the header in the lower rpm ranges. Using some exhaust port adapters for the sprint car-type spread-port flange, and a 2-inch-diameter primary tube with a 28-inch length, we probably would have solved the radical power bump between 3500 and 4000 rpm. But, trying the crossover helped point us to the culprit and told us that everything else in relation to the intake tract and the camshaft were alright. It was a very revealing lesson in engine dynamics. When you run a big-cube small-block with a bunch of cam, be prepared to run a big-diameter header primary tube.

This wraps up the "Outer Limits" 427 small-block. With damn close to 540 lbs/ft of torque and 550 horsepower, this small-block treads where most street big-blocks can't. We admit that this engine is a bit at the outer limits of what would be considered streetable, especially in terms of cost, compression,

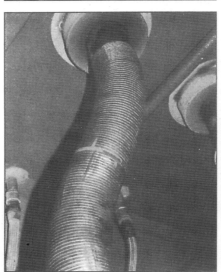

The collector crossover pipe is 3.5 inches in diameter and required a bit of work to clear all the obstacles the dyno presented. Next up was the mismatch in the flexible steel exhaust outlet tubing length and the collector-to-tubing adapter. After some re-engineering, the crossover fit and all the tubing was hooked up. And was it worth it. See the text!

There are a couple of things to note here. First is the color of the inside of the collector; it is a light tan color, meaning the carb is jetted almost perfectly for best power. Second up is the 3.5-inch-diameter collector and the 1.875-inch-diameter header primary tubes, the biggest available with the standard Chevy exhaust bolt pattern.

The final shot of the "Outer Limits" 427 small-block. With nearly 540 lbs/ft of torque and almost 550 horsepower, this small-block goes where many big-blocks can't. Not bad for a streetable small-block Chevy whose basic design started out 35 years ago at 265 cubes.

Robert Jung (seated) and Matt Compton just after we ran the 2-inch spacer on the dyno. Robert is writing down some notes on the test as he does after each and every dyno pull, while Matt is waiting for the huge electric fan to cool the dyno room off so he can make the collector crossover.

and camshaft, but that's why it was built, to test the outer limits of a street small-block Chevy. And test we did. There are over 50 runs on this engine, with some of them the hardest we've ever seen, and when we did a leakdown at the conclusion of testing, we were between 3% and 7%, the same figure you'd expect from a fresh engine. It proves that with proper parts selection and correct assembly, a small-block Chevy can be turned into a giant killer, even if it is at the outer limits.

THANKS...

A special thanks goes out to every company that participated in this project. Without them, the monster power of our Bow Tie little block wouldn't have come about. The crew at Edelbrock has to be one of the best in the world to work with, all the way from top. I want to express my heartfelt thanks and appreciation for a job well done to all, but especially to Robert Jung and Matt Compton, for whom there was no obstacle that would stand in the way of getting this testing done and done right.

Thanks!

—Jim Losee

CAMSHAFT SPECS: OLD vs. NEW

	Old Intake/Exhaust	New Intake/Exhaust
Duration at 0.050-inch (degrees)	254/260	250/252
Valve lift (in.)	0.621/0.624	0.613/0.618
Lobe separation	108	112
Valve lash-hot	0.024/0.026	0.018/0.020

RACE ENGINE BASICS

Parts And Pieces Needed To Build Up A Bracket Race Engine
By Chuck Schifsky

Car crafters love horsepower. And high horsepower generates low timeslips at the dragstrip. But with big horsepower on tap, daily driveability is usually affected. For most of us, when building an engine we are forced to choose between mild manners on the street or potent power at the strip. Usually, the mild-mannered engine takes priority. But what if you wanted to build an engine specifically for use at the dragstrip? We're not talking about a potent street/strip engine, we're talking about an engine specifically built to run hard during weekend bracket racing—every weekend throughout the racing season.

Such a bracket racing engine sees a hard life. Every time it is started up it's pushed to its rpm maximum. Pass after pass of 7000rpm shifts has a habit of destroying engines, unless the powerplant has been properly fortified with race-engineered parts. That's where this story comes in. We show you a healthy sampling of parts and pieces needed to build up a true race-only engine. For a street car many of these parts would be considered overkill, but for a race engine the heavy-duty parts are mandatory. In addition, we give you some of the key specs and clearances needed to build a race engine—to make sure everything works in harmony the first time the loud pedal is mashed on the starting line. Remember, this isn't a 100-percent start-to-finish buildup of a race engine; it's a smattering of tech tidbits along with the rationalization for choosing specific parts. We follow along as the folks at Wood Engineered build up different aspects of one of its small-block Chevrolet bracket race engines.

Building a bulletproof bottom end is critical for a competitive bracket racing engine. Simply installing a reground crank with stock bearings won't cut it. In addition, a quality set of rods and lightweight pistons are mandatory. These rods, pistons and bearings *(shown in photo)*, along with a race-prepped crankshaft, are part of a Super Bracket Kit from Lunati. The steel rods (5.7-inch length) have been upgraded with ARP bolts, feature floating pins and have been shot-peened and Magnafluxed. The lightweight (491 grams) flat-top Lunati/Taylor pistons are forged aluminum units that can handle the rigors of extended rpm and extreme cylinder pressure.

This 3.75-inch-stroke crankshaft from Lunati (part of a complete Super Bracket bottom end kit) has been fully machined and internally balanced to racing specs. Once the crankshaft is in, be sure to measure for proper endplay—Wood Engineered prefers about 0.007-inch endplay. Wood Engineered feels that if endplay is 0.004 inch then clearances are too tight, while 0.010 inch is a bit too loose. In addition, measure *each and every* crank journal (main and rod) along with the bearing thicknesses to ensure proper clearance. On this particular engine, the mains have 0.003-inch clearance while the rods have 0.0025-inch clearance.

Race Engine Basics

Use a dial bore gauge to check the size and roundness of each bore. Shoddy workmanship and/or improperly calibrated machinery can lead to inconsistent bore roundness that will affect piston-to-bore sealing. In addition, out-of-round cylinders promote increased friction (which hampers horsepower output and leads to premature wear). For race engines, we recommend using torque plates when boring and honing to keep the cylinders round under actual operating conditions.

Before installing the pistons, check the ring end gap. Usually, the ring end gap will need to be adjusted via filing. To properly check end gap requires positioning the ring(s) "square" within the bore, just as if the rings were mounted on the piston. To do this by sight is nearly impossible, thus you should use a ring squaring tool such as those sold by Tavia. Set the squaring tool in the top of the bore and then butt the ring against the bottom of the tool. Then, remove the tool and check (with a feeler gauge) the amount of ring end gap. On this particular race engine, Wood Engineered filed the Childs & Albert rings to gain a 0.018-inch end gap on the top ring and a 0.016-inch end gap on the second ring.

When inserting the piston/rod/ring combo into the cylinder bore, be extremely careful not to damage the piston rings or scuff the bore during installation. One-size-fits-all ring compressors may work, but a dedicated-diameter tapered ring compressor (such as from Childs & Albert) works best. Simply tap the piston downward, and as the compression tool tapers, the rings will compress to the desired bore diameter.

The oiling system is the lifeblood of any engine. Without proper oil pressure and volume, an engine will self-destruct. Because of the high rpm a bracket racing engine sees, in addition to the sloshing effect that occurs on each launch, oil pump and pickup placement must be spot-on. Use a level and a straight-edge ruler (or similar device) to make sure the oil pump pickup sits level and at the correct depth within the oil pan. If not, the pickup may not submerge properly in the oil within the oil pan, and big problems will result. Also, take note that if the pickup is too close to the bottom of the oil pan, it may not be able to pick up the oil correctly either.

On any race engine it's a good idea to install a windage tray. This special tray from Stef's helps keep the in-pan oil from splashing against the crankshaft as it rotates during operation. In addition, the tray helps to keep the oil from sloshing around in the oil pan, which could starve the oil pump pickup. Installing a windage tray is simple, but requires special mounting studs that are usually included in the kit.

Once you've got your oiling system figured out, don't stop short by installing a cheapo or stock oil pan. Because of harsh driving conditions (such as during lofty wheels-up launches) and severe engine rpm, installing a proper oil pan is just as important as having a quality oiling system (pump and pickup). This oil pan from Stef's Fabrication Specialties is custom-made to fit within the tight confines of the race car-to-be's engine bay. In addition, it has built-in baffles (to control oil flow) and holds extra oil (6 quarts total) to ensure that sufficient oil is available to lubricate the engine. Constructed of aluminum, this Stef's oil pan is lighter and dissipates heat better than traditional O.E.M. oil pans. Plus, if you need extra room to clear a brace or a starter, an oil pan such as this can be custom-made to your specs.

Unfortunately, to make high horsepower usually requires high-compression pistons. And, as we all know, high-compression engines wreak havoc on stock starters. Even if a stock starter can roll a race engine over, chances are the engine may be difficult to start, because the starter can't do so at a fast enough rpm. Special aftermarket starters, such as this unit from Tilton, feature internals that deliver torque multiplication that will easily turn over extremely high-compression engines under the harshest of high-heat conditions. Plus, this starter is smaller and lighter than traditional O.E.M. units, which is a blessing when working within the tight confines of a race car.

Traditional timing chain setups rotate the camshaft as dictated by the crankshaft. However, regular timing chains don't offer much tunability and are noted for transferring harmonics from the crankshaft to the cam and valvetrain. Poor harmonics rob horsepower and lead to durability problems—especially at the high rpm that race engines are subjected to. Belt-drive timing systems, such as those from Edelbrock, allow an engine builder to tune an engine for maximum horsepower or for prevailing racing conditions. The billet-aluminum upper gear (with adjustable thrust assembly) can be rotated in various stages to advance, retard or "degree" a camshaft as the engine builder desires. The Edelbrock design, with its nine-bolt vernier system, allows adjustment within $7/10$ degree for highly accurate timing. Plus, the belt design doesn't transfer crankshaft harmonics nearly as much as chains do, promoting added reliability and more accurate valvetrain operation.

Belt-drive assemblies usually go onto the cylinder block without problems, but block casting configurations and dimensions can vary from year to year. If the block casting interferes with the belt-drive cover, you'll need to carefully grind metal away as shown (arrow).

Measurements

Although aftermarket high-performance engine parts specs are usually spot-on, *always* double check. Start by checking true top dead center (according to the piston) and compare it with that indicated on the harmonic dampener. Sometimes the timing gears may be mismarked, mismachined or installed incorrectly, which will give a false indication of top dead center. Then again, the harmonic dampener may be mismarked or may have spun slightly—both of which will give a false reading of TDC. After correctly locating TDC, check camshaft phasing as well as lobe lift.

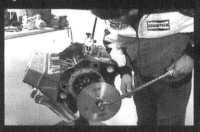

Using a magnetic-based dial indicator such as those Tavia offers, determine (and double check) the gross lift and duration on the camshaft. This ensures that the cam is both correctly machined and correctly installed. This mechanical roller camshaft from Lunati was perfect. It measured 0.662-inch lift intake and 0.624-inch lift exhaust, along with 262 degrees duration (intake) and 271 degrees duration (exhaust) at 0.050-inch lift. Avoiding mechanical interference also requires checking the piston-to-valve clearance.

Determine true compression ratio by checking the final volume of the cylinder while the piston is at TDC. Use a compression ratio tool (PN 62300 from Moroso) and a burette to check the volume of each cylinder. Compare the figures (measured in cc's) with that of the cylinder head combustion chamber and thickness of head gasket to be used.

On a bracket racing engine, airflow is everything. You could install an awesome bottom end along with a big-lift roller camshaft, but if the cylinder heads can't flow enough air, you'll be severely down on horsepower. Investing in a quality set of cylinder heads, such as these from Brodix, will enable you to release the full power potential of an engine. Made of aluminum, these Brodix Track 1 heads feature 2.08-inch intake and 1.60-inch exhaust valves paired with 1.55-inch-diameter chrome-silicon valvesprings. At a 1.995-inch installed height, the springs provide 200 psi of closed pressure. Without the proper springs matched to the camshaft, the engine will experience valve float at high rpm. Notice the large intake runners. Most serious race engine builders prefer to port-match the runners to the intake manifold for optimum flow characteristics, but these Brodix Track 1 heads come completely race-prepped—gasket-matched with blended bowls and matched runners.

Looking at the combustion chambers on a Brodix Track 1 head, you'll notice the revised combustion chambers (compared with a stock head), along with the angle placement of the spark plugs. The angled plug points toward the exhaust valve to aid in flame propagation. The combustion chamber size on these heads is 69 cc.

As with any performance engine, torquing the head bolts progressively to the proper spec is critical. In addition, due to the extreme conditions such as high rpm and cylinder pressures (especially when using nitrous oxide and/or superchargers), quality head gaskets and head bolts are mandatory. Wood Engineered uses Fel-Pro gaskets and ARP fasteners throughout its engine buildups. Notice the markings on the Champion C59C spark plugs. Each line on the spark plug denotes the orientation of the electrode tip to aid in spark plug phasing. Simply look at the line and you know where the tip is facing. Most engine builders prefer to have the spark plug fire toward the exhaust valve, claiming it helps to ignite the air/fuel mixture better.

RACE ENGINE BASICS

Usually, to make big horsepower requires spinning the engine at high rpm. At high rpm, a traditional valvetrain takes a beating, often resulting in parts failure. On bracket racing engines, it's an excellent (almost necessary) idea to upgrade to a set of full-roller rocker arms. These rocker arms can handle the rigors of high rpm and with their rollerized parts, they reduce valvetrain friction, freeing up additional horsepower. Stud girdles, such as this unit from Moroso, keep the valvetrain stable under high-rpm engine operation. On a high-output, roller camshaft-equipped engine such as this, establishing correct valvetrain geometry and maintaining proper valve lash is key to performing during high rpm operation.

Cooling a high-compression race engine is hard work. To ensure that cooling occurs properly and consistently, many racers opt for a high-flow aluminum water pump such as this unit from Edelbrock. The water pump is lighter, dissipates heat faster, and won't cavitate coolant as most stock water pumps will. In addition, this Edelbrock water pump is driven by a separate electric motor that drives the pump at a constant rpm—promoting more consistent cooling. Using a water pump drive kit (such as this unit from Moroso) also frees up extra horsepower that is normally robbed when using the traditional belt and pulley system.

Accurate ignition timing is a critical link to making high horsepower. Standard ignition systems do a decent job of delivering spark when it's needed, but crank-trigger ignition systems (such as those from MSD) are much more accurate. Why is spark timing critical? Extremely accurate timing is critical for engines that use excessively high compression ratios (above 13:1), a supercharger and/or nitrous oxide injection, where a 1- or 2-degree variance in ignition timing will lead to detonation. Crank-trigger ignition systems are mandatory for those race engines that are right on the "edge."

Installing a crank-trigger ignition system is fairly basic, but certain steps must be followed. Bolting the crank-trigger wheel onto the crankshaft is easy, but must be done so the clockwise arrow of rotation faces outward (toward you). Then install the crank-trigger pickup and bracket onto the cylinder block as shown. For the crank-trigger to work properly, there must be approximately a 0.050-inch air gap between the pickup and the trigger wheel. If, for example, the air gap is wider than 0.050 inch, the engine probably won't have a strong enough ignition signal at low rpm, resulting in difficult engine start-up.

For serious bracket racing engines, getting enough air/fuel mixture into the cylinders requires a high-flow carburetor like this unit from Barry Grant Fuel Systems. The carb started life as a 750cfm unit (PN 4779S3), but it has been reworked to flow a whopping 1040 cfm. The BG carb is perched atop a Brodix SP1 aluminum intake manifold. This combo will be able to effectively feed the needs of the Lunati 406 engine it's paired with as it operates in the 7000rpm range.

Venting unwanted exhaust gases on a high-output race engine requires special large-diameter headers. Traditional street headers will usually be too restrictive, so many racers opt for adjustable headers like those from Hooker Industries. Bolting on the special adapter plate allows racers to use big tube headers (such as 1⅞ or 2 inches in diameter for small-block use) without having the header neck down to mate with the smaller exhaust port.

Here is an example of a large tube header installed on a 406-cubic-inch bracket race engine. Notice how much larger the tubes are as compared with those on a traditional street engine. Having each exhaust runner equalized in length also delivers higher (and more consistent) horsepower. These headers from Hooker Industries even feature a special thermal coating that reduces underhood heat and improves exhaust gas flow.

Sources

Automotive Racing Products
Dept. CC
250 Quail Ct.
Santa Paula, CA 93060
805/525-5152

Barry Grant Fuel Systems
Dept. CC
RR 1, Box 1900
Dahlonega, GA 30533
706/864-8544

Brodix
Dept. CC
301 Maple St.
Mena, AR 71953
501/394-1075

Childs & Albert
Dept. CC
24849 Anza Dr.
Valencia, CA 91355
805/295-1900

Edelbrock Corp.
Dept. CC
2700 California St.
Torrance, CA 90503
310/782-2900

Federal-Mogul Corp/Speed-Pro
Dept. CC
26555 Northwestern Hwy.
Southfield, MI 48034
800/237-9090

Fel-Pro
Dept. CC
7450 N. McCormick Blvd.
Skokie, IL 60076
708/674-7700

Hooker Industries
Dept. CC
1024 W. Brooks St.
Ontario, CA 91762
909/983-5871

Lunati Cams & Equipment
Dept. CC
P.O. Box 18021
Memphis, TN 38181
901/365-0950

Moroso Performance Products
Dept. CC
80 Carter Dr.
Guilford, CT 06437
203/453-6571

MSD Ignition
(Autotronic Controls Corp.)
Dept. CC
1490 Henry Brennan Dr.
El Paso, TX 79936
915/857-5200

Stef's Fabrication Specialties
Dept. CC
699 Cross St.
Lakewood, NJ 08701
908/367-8700

Tavia Machine
Dept. CC
12851 Western Ave., Unit D
Garden Grove, CA 92641
714/892-4057

Tilton Engineering
Dept. CC
25 Easy St.
P.O. Box 1787
Buellton, CA 93247
805/688-2353

Vibratech/Fluidampr
Dept. CC
537 E. Delavan Ave.
Buffalo, NY 14211
716/895-5404

Wood Engineered
Dept. CC
537 F Constitution Ave.
Camarillo, CA 93012
805/987-0027

Using many of the tips and tricks outlined earlier, we built up a bracket racing engine using a Lunati Super Bracket Kit (406-cubic-inch Chevrolet). Even though the crankshaft is a cast design, Dale Browning (general manager at Lunati) says, "as long as you stay under 7000 rpm, the crankshaft is plenty strong enough for most bracket racers." Other key engine parts used in our buildup include a Lunati mechanical roller camshaft, Brodix Track 1 cylinder heads, a Brodix SP1 intake manifold, a Barry Grant carburetor, an MSD crank-trigger ignition system, a Fluidampr balancer and Hooker headers. This engine is destined for a 3500lb car using a TH350 automatic transmission that has a fortified chassis including a 4.10:1-geared 12-bolt rearend, ladder bar suspension and slicks. This car is fairly representative (and similar) to many of the cars that run in traditional bracket racing series' across the country. Our intent was to run the engine on a dynamometer to show what kind of horsepower our collection of parts generates, but time was short and we had to adhere to our deadlines. Don't fret, we did plug all of the vital information (including parts, cam specs, compression ratio, and so on) into a new computer program called DeskTop Dyno and came up with a set of horsepower numbers that should be darn close. Also, listed below is a rundown of the parts and pieces we used in our engine buildup. If you are considering (or are in the midst of) a similar engine buildup, you'll probably have good results using many or all of the parts we used.

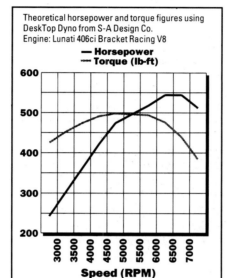

Theoretical horsepower and torque figures using DeskTop Dyno from S-A Design Co.
Engine: Lunati 406ci Bracket Racing V8

Our block was in fairly good shape, only needing to be cleaned up and bored 0.030 inch over. Installing the Lunati/Taylor forged aluminum pistons was straightforward, but keen attention had to be given to making sure we didn't scuff the cylinder walls or break a ring as the rod/piston/ring combo went in.

Our Barry Grant carburetor was a perfect match for the Brodix SP1 aluminum intake manifold. As healthy as our Lunati 406 bracket racing engine was, we needed a high-flowing, race-proven carb/intake combo such as this. In addition, we installed a BG inline fuel pressure gauge to keep tabs on fuel pump output.

406 Chevy Bracket Race Engine Components

Lunati Bottom End Kit
- Lunati custom-ground crankshaft (4.155-inch stroke)
- Lunati blueprinted 5.7-inch rods (shot-peened, Magnafluxed, new rod bolts)
- Lunati/Taylor forged aluminum pistons (with new wristpins)
- Michigan Clevite-77 rod and main bearings
- Sealed Power moly piston rings

**All parts are precision machined and balanced

Buildup Parts
- ARP bolts; main, head and intake (PN 134-5001, 134-3701, 434-2101)
- Brodix Track 1 aluminum cylinder heads
- Brodix SP1 aluminum intake manifold
- Barry Grant 04779S3 750cfm carburetor (reworked by BG to flow 1040 cfm)
- Champion C59C spark plugs
- Edelbrock camshaft belt-drive (PN 1800)
- Edelbrock belt-drive cover (PN 1810)
- Edelbrock Victor Series aluminum water pump (PN 8810)
- Edelbrock water pump spacer kit (PN 8830)
- Fel-Pro intake, oil pan, valve cover and head gaskets (PN 1206, 1821, 1604, 1014)
- Fluidampr 7¼-inch balancer (PN 712430)
- Hooker adjustable headers
- Lunati mechanical roller camshaft (PN 50148)
- Made For You Products spark plug wire holders
- Moroso electric water pump drive kit (PN 63750)
- Moroso stud girdle (PN 67070)
- MSD crank-trigger (PN 8610)
- MSD low profile distributor (PN 84697)
- MSD-7AL2 ignition (PN 7220)
- MSD Heli-Core spark plug wire set (PN 3120)
- Stef's aluminum oil pan with windage tray
- Tilton Super Street starter (PN 54-101)

We found this useful computer program that allows you to (theoretically) build and test almost any engine. The DeskTop Dyno (PN 40) has pull-down menus to make component selections a snap. Assemble an engine from millions of possible combinations quickly and easily. Then "run" the engine through a powerful full-cycle dyno simulation and watch the DeskTop Dyno produce very accurate horsepower and torque curves. Change cams, heads, compression, induction, manifolds and displacement—then run back-to-back tests. The software runs on any IBM-compatible PC. Contact S-A Design Co., Dept. CC, 515 W. Lambert, Bldg. E, Brea, CA 92621, 714/529-7999.

When doing an engine buildup, it pays to have the correct tools to aid installation and measurement of parts going into the engine. We looked through the Tavia catalog and came up with a few cool tools that you may wish to consider when doing your next engine buildup. The photo shows one such "cool tool" being used—a camshaft installation and removal tool (PN 08010).
- Tavia dial indicator (PN 08055)
- Tavia 0-6-inch dial caliper (PN 08056)
- Tavia Pro Series oil pump primer (PN 08300)
- Tavia dial bore gauge (PN 09250)
- Tavia crankshaft sockets and turning nuts (PN 08600)

SUPER MOUSE 402 CHEVY

Able To Leap Tall Big-Blocks In A Single Bound!

By Jim Losee

We started out with a Chevrolet Bow Tie block featuring siamese bores and extra-thick cylinder walls. The block was ground by Robert Jung of Alcohol Concepts. He removed the sharp edges and enlarged the oil drain-back holes to aid moving the oil back to the pan as quickly as possible. Jay Steele of Taylor Engines performed all the block machining, including the boring and honing of the cylinders with torque plates to a 4.129-inch diameter and milling the deck surface. Jay then align-honed the main journals to the requisite 350 size and installed the head studs.

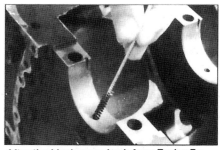

After the block came back from Taylor Engines, Robert sent it down to G&L Coatings to have the cylinder bores and lifter bores coated with their dry film lubricant to reduce friction between the piston and cylinder wall. Engine-cleaning brushes from Moroso were used to remove any debris.

Big-block Chevy. The words conjure up thoughts of awesome amounts of horsepower and torque with huge doses of cubic inches. Even thinking about making that kind of power with a street-driven small-block has the guys down at the cruise-in laughing. Well, not any more. We'll show you how to build one using some common sense and good engine-building techniques, a small-block Chevy, and off-the-shelf parts.

In the November '89 issue of CAR CRAFT we presented the results from the flowbench test of the stock Chevy head (76cc chamber, straight plug, 2.02/1.60-inch valve diameter, 163cc intake port volume, cast iron head) and the Dart II cast-iron head (72cc chamber, angle plug, 2.02/1.60-inch valve diameter, 180cc and 200cc [the figure 220cc appeared erroneously in the article] intake port volume). For a mild street application the Chevy head was the winner, but at the higher end of the street performance spectrum, the Dart II 180 head was the answer. For all-out competition applications, the Dart II 200 head worked best.

Because most readers drive street machines with some performance modifications, we thought it would be appropriate to test the Chevy and the Dart II heads head-to-head on the dyno. To find out just what type of engine combination it would take, we talked to Robert Jung of Alcohol Concepts in Los Angeles about what he would recommend as a reliable test engine. We told him we had to have an engine that would be a big enough air pump to effectively test the heads through the whole rpm range and last, too.

To that end, Robert came up with the 402cid with a few angles that aren't usually done in a street engine of this size. After some coaxing, Robert agreed to assemble the 402 for us, too. One of the angles he alluded to was the use of a 6-inch center-to-center connecting rod to help improve the rod-to-stroke ratio. With the stock 400 Chevy rod (5.565 inches center-to-center) and crank (3.750-inch stroke) combination, the ratio is

SUPER MOUSE 402 CHEVY

Upon trial assembly, we found that the Competition Cams' 6.00-inch center-to-center rods hit the pan rail on every cylinder in the block. After several hours worth of grinding, all the rods cleared the block. So the rods wouldn't catch the Fel-Pro pan gasket and pull it out, the pan gaskets were held down using ARP's pan bolts and notched for rod bolt clearance.

Because of the long stroke and rod length, the wristpin hole intersects the oil control ring on these pistons. To help aid oil control ring strength, Wiseco cut the oil ring groove for a support spacer. Speed Pro came up with the ring set that consisted of a moly top compression ring that was hand-fitted to each bore.

1.48:1. Using the same 3.750-inch stroke and the 6-inch center-to-center Comp Cams rods, the ratio goes up to 1.6:1. The higher the ratio (up to a certain point, due to physical limitations of the block, cam, rods, pistons, and crank), the more power can be made.

Getting this reciprocating assembly combination in the Bow Tie

In order to withstand the rigors of the dyno-testing, a Cola Performance non-twisted 4340 forged steel crankshaft was selected. The crank has a very generous radius on both the main and rod journal fillets that help increase the strength. To help longevity and to ensure continuous lubrication, G&L Coatings applied their dry-film lubricant on all the bearing surfaces of the crank, as well as on all the Speed-Pro main and rod bearings themselves.

Finding an off-the-shelf piston for the 4.129 bore and the 6-inch Competition Cams rods wasn't easy, but Wiseco and Bill Mitchell Hard Core Racing Parts came up with an octet. To aid flame travel and combustion efficiency while trying to keep the compression down, Alcohol Concepts had Mitchell remove the dome and make these flat-toppers. Wiseco usually makes these pistons for Sprint car engines, so they have extremely strong pin bosses and skirts to withstand serious stress. As before, we coated the pistons with dry-film lubricant to fight friction.

Timing accuracy is critical in making power. Connecting the crank and the cam with the Donovan Gear Drive ensures that we got the most accurate timing possible. In order to install the Donovan Gear Drive crank gear, Robert had to hone it 0.002-inch for a 0.003-inch press-fit on the crank snout. The only way the gear can be installed is by using a brass drift and gently tapping it. Robert also degreed-in the Crane roller cam and it came out to the same specs as on the cam card.

Chevy block required that the pan-rails on the block be notched for clearance. It took Robert several hours and a couple of trial fittings to get both the rods and the counterweights of the Cola Performance crank to clear the inside of the block.

Cola Performance made the crank for the 402. It is an absolutely flawless piece of workmanship and measured perfectly in all dimensions. There are a couple of things that need to be mentioned when using a crank like this for one of these engines. First, because the Bow Tie block is machined for the 350 main journal size of 2.45 inches and not the standard 400's 2.65-inch journal, the crank had to be a heavy-duty 4340 non-twisted forging in order to withstand the forces exerted by a 3.750-inch stroke. To retain the torsional strength while using the 350 main size, Cola Performance increased the radii on the throws and main journals, with which consequently a standard-width bearing won't fit. But, with a little work, a chamfer was put on the bearings and they fit like a charm and cleared easily. And we now know that there are bearings on the market that are made this way already.

The second item in relationship to the crank is the balancing. According to Automotive Balancing

After the main bearing clearances were checked (0.002 inch) the ARP maincap studs were torqued down for the last time. The ARP studs used by Alcohol Concepts are the extended type that allow the use of a windage tray. The clearances on the rod bearings were checked and found to be 0.002 inch, which was within specs, so the Comp Cams rods were torqued down using ARP's super premium 2000-series rod bolts.

A complete wet-sump oiling system from Moroso was used, including this special high-volume pump with the specially machined reliefs and transfer slots. The extra machining done to the pump prevents cavitation and spark scatter under heavy loads. Even though the pick-up to the oil pump bolts directly to the pump body, we had Tony Ruthven of Ruthven Engineering TIG-weld the pick-up to the pump to be doubly sure.

Service, there are two ways to go with balancing this combination. The first involves using a stock 400-style harmonic damper and flywheel which are externally balanced, meaning they have counterweights built into them. Using this combination could prove disastrous, as the stock 400 damper wasn't designed for high-rpm operation and it can be hard on the snout of the crank. The second way is more involved and much more costly, but is more reliable and provides better dampening in the environs that this 402 engine was going to see. It entails using a Fluidampr damper for a 350cid engine and a standard 350-type steel flywheel from McLeod, and also using Mallory metal on the crankshaft counterweights to in-

INTERSTATE offers a complete line of Cycle-Tron® batteries built to withstand the toughest use in motorcycles, jet skis, ATVs, lawn and garden equipment and more.

CALL 1-800-CRANK IT FOR A DEALER NEAR YOU.

SUPER MOUSE 402 CHEVY

ternally balance the 402 like a regular 350.

The only marginally streetable part of this engine is the large cam sizing. This engine doesn't exhibit

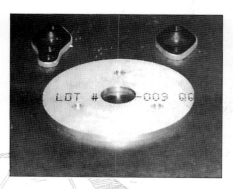

Robert checked piston-to-valve clearance and found there was over 0.240 inch of clearance on both the Manley intake and exhaust valves. The Wiseco pistons were 0.027 inch down in the bore, which helped both the valve clearance and kept the compression down to a manageable 9.8:1 with the Chevy heads and 10:1:1 with the Dart II heads.

Keeping the oil away from the crank was done using a Moroso windage tray on the adjustable-height ARP main studs. The reason behind the windage tray is that it will prevent the oil from sloshing onto the crank and taking power away, and it prevents the oil from foaming and heating up. A Moroso 7-quart extended side sump pan covered the bottom end. In order to clear the long-stroke crank-and-rod combination the pan had to be dimpled in a couple of spots.

Because the back of the factory short-leg water pump hit the front of the Donovan Gear Drive cover, Robert had to have some spacers made for the water pump and the damper pulley. The reason for the damper pulley spacer was to ensure that the water pump pulley and the damper pulley would line up. Paul Smithsky of Edelbrock built our ¼-inch spacers.

To aid header clearance and expedite the oil flow to the cooler, we used a Traco oil filter adapter/bypass. A pair of Hedman's finest headers were used featuring a 1.75-inch primary tube diameter with a 3-inch collector and 32-inch-long tubes.

One of the most versatile intake manifolds on the market today for the small-block Chevy is the Edelbrock Victor Jr. (PN 2975). The Victor Jr. has an rpm range from about 2500 to 7000 and will make power all the way. The manifold used a thermal barrier on the bottom to keep oil heat away, as well as a protective finish on the exterior surfaces, all applied by G&L Coatings. The valvetrain is shrouded by a pair of Moroso crinkle black die-cast aluminum circle track covers with twin breathers on the left cover to vent the engine.

There are two things to note here. First and most important, is the use of a damper installation tool. There is no excuse for using a piece of wood and a hammer to install a damper anymore. Secondly, we used a Fluidampr harmonic dampening device and 30-pound McLeod steel flywheel which were balanced by Automotive Balancing Service.

Checking the clearance between the ARP 7/16-inch rocker studs and the bottom of the Crane 1.5:1 aluminum roller rocker arms showed that the rockers hit due to the lift geometry of the 0.625-inch lift cam. To cure the problem, Alcohol Concepts got some Crane pushrods that were 0.100-inch longer than stock and this cured the problem. Sealing the heads and all the other gasket surfaces on the 402 was done utilizing Fel-Pro gaskets, most notably the PermaTorque/Blue head gaskets and the Printoseal intake manifold gasket.

Carburetion comes in the form of a Holley PN 4779 750cfm carb given the full-on Saturday Night Special treatment by Fuel Curve Engineering of Tryon, North Carolina. Providing spark is a Mallory Comp 9000 electronic distributor and coil with Moroso Blue Max wires and separators sending the fire to the spark plugs.

SUPER MOUSE 402 CHEVY

the idle qualities we like to see in a street performance cam, but it really helped to see what the airflow/power characteristics of both sets of heads were. And if a rumpity-rump idle is what you want, this Crane cam will deliver. But, oh what power it yielded in the select-ed rpm range. The accompanying photos and captions detail the buildup of the 402 Super Mouse motor and the final results of the heads-up head-to-head shootout.

....AND THE ENVELOPE PLEASE

The heads tested here were randomly selected and used right out of the box except for the valve job done by Port Flow Design. No port matching or bowl work was done to the heads. The Edelbrock Victor Jr. manifold we used was right out of the box without modifications. Engine timing was set at 34 degrees total. This was the point at which we got the best power numbers using 104 octane racing gasoline.

And now the results:

and valvetrain components are about $800 on the open market, whereas the Chevy heads list out at $690 a pair bare.

It's true that this wasn't exactly an apples versus apples test, as the Dart II heads had angled spark plugs and smaller combustion chambers that yielded 0.3 more compression, but that still doesn't equate to such a dramatic increase in power. The answer is that the Dart II heads work very well right out of the box.

STOCK CHEVY HEAD TEST

RPM	Corrected Torque	Corrected Horsepower	Volumetric Efficiency	Brake Specific Fuel Consumption
2500	403.2	191.9	89.7%	0.49 lb/HPhr
3000	402.4	229.9	87.7%	0.48 lb/HPhr
3500	406.4	270.8	87.9%	0.49 lb/HPhr
4000	406.1	309.3	89.3%	0.48 lb/HPhr
4500	411.9	352.9	93.3%	0.50 lb/HPhr
5000	416.5	396.5	97.1%	0.51 lb/HPhr
5500	394.8	413.4	94.4%	0.52 lb/HPhr
6000	356.0	406.7	89.9%	0.55 lb/HPhr

Chevrolet heads: 2.02/1.60 valves, 76cc chambers, 163cc intake runners, straight plug, cast iron

DART II HEAD TEST

RPM	Corrected Torque	Corrected Horsepower	Volumetric Efficiency	Brake Specific Fuel Consumption
2500	404.8	192.7	89.0%	0.47 lb/HPhr
3000	426.7	243.7	89.4%	0.47 lb/HPhr
3500	431.7	287.7	90.4%	0.46 lb/HPhr
4000	431.6	328.7	92.0%	0.47 lb/HPhr
4500	448.8	384.5	97.3%	0.47 lb/HPhr
5000	456.2	434.5	101.8%	0.48 lb/HPhr
5500	434.7	455.2	99.8%	0.50 lb/HPhr
6000	396.9	453.4	96.2%	0.53 lb/HPhr

Dart II heads: 2.02/1.60 valves, 72cc chambers, 180cc intake runners, angle plug, cast iron

Out of the box, the Dart II heads are worth almost 40 lbs-ft of torque and almost 42 hp over the Chevy heads! This is a bolt-on with no work done to the heads. As we said in the previous article, a pair of the Dart II 180s complete with valves

By the time you read this we will have taken the Dart IIs back to Port Flow Design for some port matching and bowl work to see whether we can increase both the torque and horsepower. We'll keep you informed of the results.

SOURCES:

Automotive Balancing Service
Dept. CC
P.O. Box 1984
South Gate, CA 90280
213/564-6846

Automotive Racing Products
Dept. CC
8565 Canoga Ave.
Canoga Park, CA 91304
818/341-4488 (in CA)
800/826-3045

Chevrolet
Available at over 5000 dealerships nationwide

Cola Performance
Dept. CC
19122 South Santa Fe
Rancho Dominguez, CA 90221
213/639-7700

Crane Cams Inc.
Dept. CC
530 Fentress Blvd.
Daytona Beach, FL 32014
904/258-6174

Donovan Engineering Corp.
Dept. CC
2305 Border Ave.
Torrance, CA 90501
213/320-3772

Edelbrock Corp.
Dept. CC
2700 California St.
Torrance, CA 90503
213/781-2222

Fel-Pro Inc.
Dept. CC
7450 N. McCormick Blvd.
Skokie, IL 60076
312/674-7700

Fuel Curve Engineering
Dept. CC
Rte. #2, Box 35
Tryon, NC 28782
704/894-3511

G&L Coatings
Dept. CC
888 Rancheros Dr.
Unit F2
San Marcos, CA 92069
619/743-0224

Hedman Hedders
Dept. CC
9599 West Jefferson Blvd.
Culver City, CA 90232
213/839-7581

Holley Replacement Parts Division
Dept. CC
11955 East Nine Mile Road
Warren, MI 48089
313/497-4250

McLeod Industries Inc.
Dept. CC
2906 E. Coronado
Anaheim, CA 92806
714/630-2764

Mallory Ignition
Dept. CC
550 Mallory Way
Carson City, NV 89701
702/882-6600

Manley Performance Products
Dept. CC
1960 Swarthmore Ave.
Lakewood, NJ 08701
201/905-3366

Bill Mitchell Hard Core Racing Parts
Dept. CC
80 Tradezone Ct.
Ronkonkoma, NY 11779
516/737-0372

Moroso Performance Parts Inc.
Dept. CC
80 Carter Drive
Guilford, CT 06437
203/453-6571

Port Flow Design
Dept. CC
348 "C" East Carson St.
Carson, CA 90745
213/835-4457

Speed-Pro Sealed Power Corp.
Dept. CC
100 Terrace Plaza
Muskegon, MI 49443
616/724-5011

Taylor Engines
Dept. CC
8145 Byron Rd., Unit D
Whittier, CA 90606
213/698-7231

Traco Engineering
Dept. CC
11928 West Jefferson Blvd.
Culver City, CA 90230
213/398-3722

Wiseco Piston Inc.
Dept. CC
7201 Industrial Park Blvd.
Mentor, OH 44060
216/951-6600

World Products
Dept. CC
343 Oliver
Troy, MI 48084
313/244-9822

THANKS TO:

First things first. We want to say a big thank you right up front to all the manufacturers who provided equipment and time to this project. Without their cooperation this engine wouldn't have happened and the results wouldn't have been as great as they were. Another big thanks goes to the guys at Alcohol Concepts, especially Robert Jung for worrying about the details and taking the time to put this engine together correctly. And the final thank you goes to all the folks at the Edelbrock Corporation who assisted in the dyno testing, all the way from the top to all the great guys in the dyno room, including Jack Ring, Curt Hooker, Robert Jung, and "Big" Bill Fletcher.

Wild Thang!

BRODIX STREET HEADS AND A LUNATI 408 MOUSE MOTOR TEAM UP TO BUILD OVER 475 LBS.-FT. OF TIRE-SHREDDING TORQUE!

By Jeff Smith

Every town in America has replayed the same story. The local hot-rodding hero has built the ultimate street car, and he promotes his engine-building expertise by placing a $100 bill on the dash and then stomping on the throttle, daring his unsuspecting passenger to "just try and grab the bill off the dash." Of course, the car's acceleration rate is so great (according to the story) that not even Charles Atlas can snag the guy's money.

Street mythology aside, we thought that building a big small-block Chevy torque monster, in the best Max-Torque tradition, would be enormously fun. That idea quickly evolved into a test on two sets of affordable Brodix street aluminum cylinder heads. Then we decided to add a pair of pocket-ported iron production castings to use as a baseline. The main goal would be tons of torque, using only pump gas.

THE CONCEPT

The plan started with a Lunati 408-cubic-inch, mail-order short-block kit. The kit consists of the major machined components necessary to build a streetable big-inch Mouse motor. What attracted us to the kit was its reasonable, forged-piston price, and that all we had to do was check the clearances and assemble the engine. Lunati also offers the choice of any hydraulic Lunati camshaft. We opted for the lumpy Street Master single-pattern hydraulic with an advertised 285 degrees of duration, 235 degrees of duration at .050-inch cam lift, .507-inch valve lift, and a very tight 108-degree lobe-separation angle. This close lobe-separation angle was chosen to move the torque down in the rpm range, enhancing low- and midrange power. Of course, this tight lobe-separation angle will also hurt idle quality, but we felt that the larger 408-inch engine could still idle decently, at the sacrifice of vacuum-operated accessories such as power brakes.

The choice of Brodix cylinder heads also required some thought, since Brodix offers a number of cylinder heads and options. Since most street enthusiasts will opt for the entry-level Street head (which is currently advertised for $995), it was our first choice. The Street head comes with a relatively small 165cc intake-port volume, 2.02/1.60-inch intake and exhaust valves, and is completely assembled with a valve job and pocket cleanup. It's ready to run right out of the box.

The second Brodix head we chose was also a Street head, but with an optional larger -8 intake port, which increases the port volume from the Street head's 165 cc's to approximately 185 cc's. This larger port also gets a

We started with Lunati's mail-order, .040-over, 408-cid engine kit consisting of a fully machined four-bolt main block; cast-iron crank; forged, replacement-dished Speed-Pro pistons with single moly rings; and Michigan 77 rod and main bearings. We checked all the clearances in a trial assembly to eliminate potential problems. The effort paid off when we discovered tight rod-bearing clearances on a couple of rod bearings due to a tolerance stack-up situation. This is why it always pays to double-check all machine work. The final responsibility always lies with the person who assembles the engine.

PARTS LIST	
The following is a list of the major components used in the assembly of the Lunati/Brodix 408 engine. Many of the small components not listed can be obtained through these sources.	
COMPONENT	PART NUMBER
BRODIX, INC. Street Head	N/A
Street Head w/ -8 intake	N/A
LUNATI CAMS, INC. 408 Engine Kit	N/A
Street Master 235 Cam	07102
Timing-Chain Set	93099
Rocker Arms	85025
Oil Pump	M-55
Valvesprings	73943
AUTOMOTIVE RACING PRODUCTS (ARP) Head Bolts	134-3601
Oil Pan	234-1901
Balancer Bolt	134-2501
Timing Cover	200-1401
Intake Bolts	434-2101
Valve Cover	200-7601
Other Assorted Fasteners Also Used	N/A
FEL-PRO, INC. Head Gasket	1004
Intake Gasket	1206
Exhaust Gasket	1404
Oil Pan	1802
Valve Cover	1604
Other Assorted Gaskets Also Used	N/A
FLAMING RIVER INDUSTRIES Stainless Intake Valve, 2.02 inch	2061
Stainless Exhaust Valve, 1.60 inch	2004
EDELBROCK CORPORATION Performer RPM Intake	7101
Victor Jr. Intake	2975
Valve Covers	4649
Spacer, two inch	8712
HOLLEY REPLACEMENT PARTS 750-cfm Double-Pumper Carb	0-4779
B&M AUTOMOTIVE PRODUCTS, INC. Steel-Degreed Balancer (400 style)	64266
MIDWAY INDUSTRIES Steel Flywheel, Externally Balanced	700140

The iron 441 heads were massaged slightly by Rod Sokoloski, who also did some mild bowlwork and three-angle valve jobs to both the intake and exhaust seats. The valves are stainless-steel gems from Flaming River Industries that certainly played a role in enhancing the flow potential. Rod also milled them slightly to produce 72cc combustion chambers, making the iron-head compression ratio 9.5:1. We also used taller Lunati springs on the iron heads, since production Chevy springs will coil bind with the .507-inch-lift cam!

The kit also comes with your choice of Lunati hydraulic or solid lifter camshaft. We chose Lunati's Street Master 235, consisting of a 285 advertised duration, 235 degrees at .050-inch lift, .507-inch valve lift, and ground on a 108-degree lobe-separation angle. Lunati suggested that this cam be installed advanced four degrees to a 104-intake centerline to improve the idle quality, which we also did.

larger 2.05-inch intake valve, while the exhaust port remains the same. Our test heads also came with a back-cut exhaust valve that improved the exhaust flow over the standard Street head exhaust. We added a set of Lunati springs and retainers to this head, but otherwise it was outfitted exactly as the Street head.

Both aluminum heads were ordered from Brodix with 64cc chambers, which put the compression at 10.2:1. While this may seem like an invitation to detonation, aluminum heads usually can be bumped approximately .5 to .75:1 of compression over iron heads without getting into detonation. This is due to the greater thermal conductivity of aluminum, which pulls heat out of the combustion chamber and reduces the overall cylinder pressure that could otherwise lead to detonation. The iron heads were milled to 72 cc's and came in at 9.5:1. Obviously, the comparison of the power output between aluminum and iron heads is not valid because of the difference in compression, but we included the iron heads because it is what a typical hot rodder could expect by stepping up from a good set of production heads to the Brodix aluminum heads.

For the iron pieces, we chose a set of 441 heads that had been fitted with larger 2.02/1.60-inch stainless-steel valves from Flaming River Industries. While these valves are more expensive than production pieces, they are an extremely high-quality valve. Since installing larger valves would require machine work, Rod Sokoloski, the resident cylinder-head man at Jim McFarland's new research-and-development facility in Torrance, California (see sidebar), performed a minor amount of

WILD THANG!

The basic Brodix aluminum Street head is a high-velocity intake port head with 2.02/1.60-inch intake and exhaust valves and a 165cc intake port. These heads come complete with all the springs, retainers, keepers, seals, screw-in studs, and guide plates, ready to bolt on. In fact, that's exactly how we tested them, right out of the box. The only difference between the two Brodix heads tested was the larger -8 intake port, a larger 2.05-inch intake valve, and a back-cut exhaust valve on the larger head to improve exhaust-port flow.

The entire engine was assembled with Automotive Racing Products (ARP) bolts, head bolts, intake, header, timing chain and cover, oil pan, oil-pump stud, water pump, and distributor hold-down stud. Ensuring that everything remained leak-free was Fel-Pro's job, with its complete line of head gaskets, intake, exhaust, valve cover, oil pan, and timing-chain-cover gaskets.

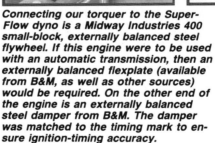

Connecting our torquer to the Super-Flow dyno is a Midway Industries 400 small-block, externally balanced steel flywheel. If this engine were to be used with an automatic transmission, then an externally balanced flexplate (available from B&M, as well as other sources) would be required. On the other end of the engine is an externally balanced steel damper from B&M. The damper was matched to the timing mark to ensure ignition-timing accuracy.

The dyno testing was done at Jim McFarland's brand-new facility in Torrance, California, by Kevin McClelland. All runs were optimized for fuel flow and corrected for atmospheric conditions. Additional components used to complete the engine included an MSD-6AT ignition, both the new Edelbrock Performer RPM dual-plane intake (seen here) and a Victor Jr., and a Holley 750-cfm double-pumper carburetor.

bowlwork to the heads to improve their flow potential. These heads are not totally optimized, but they do represent a significant airflow improvement over a set of 441 production heads.

THE TEST

Before we actually put the heads to the test, Rod put them through their paces on the SuperFlow 600 flow bench to establish flow comparisons between all three sets of heads. As you can see in flow graphs 1 and 2, there are no dramatic differences in flow, with the exception of the larger Brodix head at and above .450-inch valve lift. There are some subtle differences in flow between the three heads, with the most significant being the excellent exhaust-to-intake flow relationship of the iron 441 compared to the aluminum heads. Many engine builders feel that an exhaust port capable of over 80 percent of intake flow will make good power, especially when used with a single-pattern cam. Also, pay special attention to how each head flowed in the mid-lift areas, since they play an important role in producing torque for street engines.

We also wanted to eliminate any performance variables based on intake design, so we chose both Edelbrock's new Performer RPM dual-plane intake and the excellent Victor Jr. single-plane intake. Rounding out the package was a Holley 750-cfm double-pumper carburetor, a complete MSD-6AT ignition, and a set of Hedman 1¾-inch headers, which were run without mufflers.

With everything assembled, we contacted Jim McFarland to schedule a test of these heads on the new Super-Flow 7100 dyno. The 7100 is the next step up from the venerable 901 dyno and offers more features, including the capabilities of the new Engine Cycle Analyzer (ECA) system, which can measure the actual combustion. Dyno technician Kevin McClelland began the test with the iron 441 heads as the baseline, producing the best torque numbers with the Performer RPM manifold of 448 lbs.-ft. of torque and 383 horsepower at 5000 rpm. Predictably, the Victor Jr. traded torque for horsepower, cranking 434 lbs.-ft. of torque, but adding 10 horsepower to 393 at 5000 rpm. While these numbers were stout, the Brodix heads proved to be

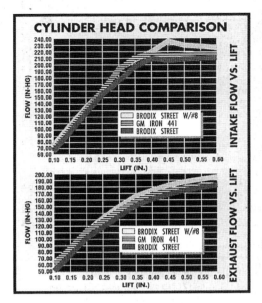

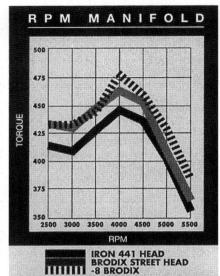

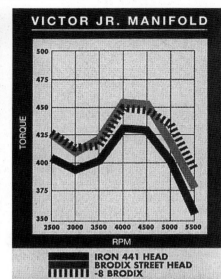

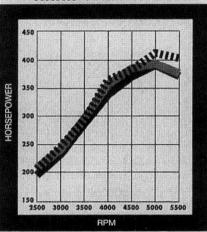

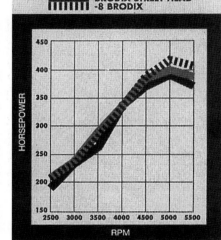

The Victor Jr. really shined in the test of the Street Brodix -8 head, where it offered not only a good balance of torque and horsepower but also fit the larger Brodix intake port much better. While we were able to just seal the smaller Performer RPM manifold to the -8 intake port, you should weld and machine additional sealing beads to the upper and lower ends of the manifold ports to effectively seal the manifold for everyday use. The Fel-Pro intake gaskets performed their sealing job well in this particular instance. The Victor Jr. was also the horsepower champ in our manifold comparison.

	RPM MANIFOLD						VICTOR JR. MANIFOLD					
SPEED RPM	TORQUE			HORSEPOWER			TORQUE			HORSEPOWER		
	IRON 441 HEAD	BRODIX STREET HEAD	-8 BRODIX	IRON 441 HEAD	BRODIX STREET HEAD	-8 BRODIX	IRON 441 HEAD	BRODIX STREET HEAD	-8 BRODIX	IRON 441 HEAD	BRODIX STREET HEAD	-8 BRODIX
2500	417.5	438.6	436.9	198.7	208.8	208.0	407.2	424.4	426.9	193.8	202.0	203.2
3000	414.4	432.8	436.3	236.7	247.2	249.2	399.1	415.4	416.9	228.0	237.3	238.1
3500	426.3	448.8	446.1	284.1	299.1	297.3	404.2	424.4	420.8	269.4	282.8	280.4
4000	448.3	466.2	476.4	341.4	355.1	362.8	434.4	454.7	455.0	330.8	346.3	346.5
4500	434.4	454.5	463.2	372.2	389.4	396.9	433.9	453.6	459.5	371.8	388.7	393.7
5000	402.8	417.4	430.9	383.5	397.4	410.2	413.1	420.8	437.8	393.3	400.6	416.8
5500	357.6	367.0	387.5	374.5	384.3	405.8	364.2	378.2	392.2	381.4	396.1	410.7

even better.

The small Brodix Street heads were next on the agenda and, with the RPM manifold, immediately proved their worth with an overall jump in the torque curve, maxing out at 466 lbs.-ft. at 4000 rpm (an 18-lbs.-ft. increase) and 397 horsepower at 5000 rpm. Switching to the Victor Jr. manifold moved the torque to 454 at 4000 and upped the horsepower to 400 at 5000 rpm. While these numbers are exceptionally strong (remember, this is using ordinary 93-octane unleaded supreme), the larger Brodix heads were still to come!

Kevin pulled the cylinder-head swap in record time, and within hours he was twisting the SuperFlow power absorber, looking for power. Again the numbers improved. Using the Performer RPM manifold, the larger Brodix heads equaled the previous test's torque on the bottom and then cranked out 10 lbs.-ft. more at peak torque to 476 at 4000 rpm while bumping peak horsepower to 410 at 5000. Even better, the bigger heads carried the power through to 5500 with 406 horsepower, a full 22 ponies over the same rpm point with the smaller heads! By this time, the Victor Jr. test was predictable, again trading torque for horsepower, and this time generating 459 lbs.-ft of peak torque at a higher 4500 rpm while making 416 horsepower at 5000.

CONCLUSION

There are a number of significant results from this dyno flog, not the least of which is the tremendous torque produced by this combination. While cubic

TESTING IN THE '90s

Every once in a while, someone steps up to do the job right. In this case, Jim McFarland has assembled a group of highly talented individuals and placed them in a state-of-the-art facility designed to push the limits of high-performance research into the future. We were lucky enough to be one of the first to be invited down to view McFarland's new facility, and were in fact the first to put his new dyno cell to the test.

While anyone can construct a dyno cell, Jim has taken that art one step further with a complete cell that is not only sound-insulated well enough so that you can barely hear an unmuffled engine run while standing less than five feet from the engine, but the air pressure inside the cell can also be controlled by dual intake and exhaust fans that circulate up to 60,000 cfm of air! The dyno itself is SuperFlow's latest SF-7100, featuring the latest in dyno technology, including the capability of measuring combustion pressure inside the chamber! Of course, experience is important, too, and the combination of Kevin McClelland's and Louis Hammel's knowledge is a big plus to this program.

But the dyno is just one phase of McFarland, Inc. The facility also has a complete cylinder-head flow department run by Rod Sokoloski, where intake-manifold, cylinder-head, and exhaust-flow parameters can be accurately tested on a state-of-the-art SuperFlow flow bench. In addition, Keith Rudofsky is the Computer Aided Design (CAD) engineer on staff who can design a component on the computer that can be quickly converted into a working prototype.

The goal of McFarland, Inc. is to offer a high-tech research-and-development facility not only to Detroit but also to any aftermarket manufacturer interested in producing leading-edge, high-performance equipment for the 1990s. It's a great way to start the decade.

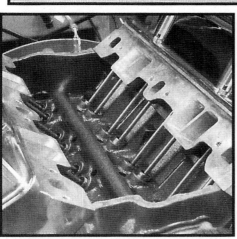

When bolting on the Brodix aluminum heads, always make sure there is sufficient clearance between the pushrods, guide plate, and the cylinder head. Often, you must compromise guide-plate location to keep the pushrod from either angling the rocker arm off the valve or hitting the intake port.

This is the McFarland team. From the left: Kevin McClelland, Rod Sokoloski, Louie Hammel, Keith Rudofsky, and Jim McFarland. There is some serious talent in this group.

inches certainly played a big role, a number of other factors also came into play—not the least of which is the amazing power potential displayed by the Brodix aluminum heads, especially the smaller Street heads whose small port volumes many considered restrictive at the beginning of this test. But if recent tests by John Lingenfelter—with big-block Chevy engines—prove anything, it's that ports once considered "too small" for a good street engine by most authorities may in fact be "just right" for strong torque.

There are a couple of warnings that we must include with this test, however. Most important, comparisons between the iron heads and either aluminum heads are flawed by the difference in compression. A total of .7:1 ratio is more than significant and could represent the difference between the Brodix Street head and the Chevy iron head. Nevertheless, we saw this as an important baseline test, since this head represents a very well-executed iron street head.

We also should point out that neither Brodix head was modified. They were tested as delivered, and offer tremendous opportunities for even more power with some minor tweaking, especially on the exhaust side. Improving the exhaust flow would offer interesting high-rpm power increases. These heads also offer other advantages over iron heads, such as removing up to 40 pounds of weight and easy repair should the heads be damaged; and certainly the image of aluminum heads doesn't hurt. They can even be ordered polished, direct from Brodix, for an additional fee.

Also note in all the dyno runs that there is an obvious dip in the torque curve between 3000 and 4000 rpm. While many theories abound, it appears that additional tuning could enhance the torque in this area. If traction were a problem, the Victor Jr. would also be a wise choice.

There is much more detailed information that can be pulled from these tests, but space limits our addressing all the combinations. Overall, the test was a dramatic success and, while the controversy over a short connecting rod versus a long rod will probably continue, it's obvious that you can still make good power even if the rod is "too short." This engine still has tremendous potential to deliver, and if enough readers are interested, we could be convinced to put this baby back on the SuperFlow dyno to optimize one set of heads—along with a massaged hydraulic cam—to see how far we can take this fully streetable, muscle-bound Mouse. Is there a Wild Thang in your future? Just say "cheese!" **HR**

SOURCES

Automotive Racing Products, Inc.
Dept. HR
250 Quail Ct.
Santa Paula, CA 93060
(805) 525-5152
(800) 826-3045

B&M Automotive Products, Inc.
Dept. HR
9152 Independence Ave.
Chatsworth, CA 91311
(818) 882-6422

Brodix, Inc.
Dept. HR
301 Maple St.
P.O. Box 1347
Mena, AR 71953
(501) 394-1075

Edelbrock Corporation
Dept. HR
2700 California St.
Torrance, CA 90503
(213) 781-2222

Fel-Pro, Inc.
Dept. HR
7450 N. McCormick Blvd.
P.O. Box 1103
Skokie, IL 60076
(312) 761-4500

Flaming River Industries
Dept. HR
34425 Lorain Rd.
Cleveland, OH 44039
(800) 648-8022

Holley Replacement Parts
Dept. HR
11955 E. Nine Mile Rd.
P.O. Box 749
Warren, MI 48090
(313) 497-4000

Lunati Cams, Inc.
Dept. HR
P.O. Box 18021
4770 Lamar Ave.
Memphis, TN 38181-0021
(901) 365-0950

McFarland, Inc.
Dept. HR
390 Amapola Ave.
Suite 1
Torrance, CA 90501
(213) 533-6100

Midway Industries
Dept. HR
11516 Adams St.
Midway City, CA 92655
(714) 898-4477

Junker to Stroker

Chevy 383 Street Torquer

BY MATTHEW KING
Photos by Matthew King

Greed for speed and cubic inches has driven generations of car crafters to stroke just about every engine imaginable, but probably the most prolific combination is a 3.75-inch-stroke Chevy 400 small-block crank in an 0.030-over 350 block netting 383 cubes.

Dropping a factory 400 crank into a 350 block has always been pretty straightforward, but until recently there were a few details and drawbacks to consider. First, the 400's 2.65-inch main journals had to be turned down to fit in the 350's 2.45-inch bearing saddles. Second, since GM never produced forged-steel cranks for the 400, you had to use a cast crank. This was fine for street use, but a cast crank wouldn't likely last long under hard-core racing abuse. Finally, due to the fact that the 400 is externally balanced, swapping in an OE crank required either the 400's weighted balancer and flywheel/flexplate or the use of expensive heavy Mallory metal for internal balancing.

However, the popularity of the conversion has resulted in a slew of aftermarket offerings that make these drawbacks distant memories, and now you can purchase drop-in cast or forged cranks to build a 383 without machining the journals or external balancing. Many companies offer rotating-assembly kits to build a 383 using your own 350 block as well as complete engine kits. In other words, building a 383 has never been easier, and one can be built to just about any budget level.

Car Craft's goal with this 383 was to build a street engine capable of neck-snapping acceleration off the line. We wanted an engine that would pull 5.0 hard from idle to a peak of around 5,000 rpm and far surpass the 400lb-ft mark, so we selected components to maximize low-end power. Starting from scratch, we turned a $100 junkyard refugee into a ground-pounding stroker for about $4,000 from carb to oil pan.

On the dyno with 92-octane pump gas, our stroker made 338 hp at 4,800 rpm and 426lb-ft of torque. Exhibiting a very flat torque curve, it thumps out a minimum of 400lb-ft from 3,000 to 4,300 rpm. Our goals were met, but there's no avoiding the fact that engine building, like life, is full of compromises. In our quest to build maximum low-end grunt, we had to sacrifice some horsepower to the torque gods. We could have put together a higher-horsepower mill for about the same money by using a bigger cam and different heads and intake, but since the engine is destined for stoplight combat, we chose to put most of our eggs in the torque basket.

A four-bolt Chevy 350 from a '75 van in a local junkyard was ours for a C-note - and didn't even have to pull it! We didn't absolutely *need* a four-bolt for the buildup, but since it cost the same hundred bucks as the two-bolt sitting next to it, we scooped it up.

Teardown revealed a virgin block with minimal ridge wear, and the block passed a Magnaflux test with flying colors. Magnafluxing also revealed the smogger heads to be junk - like we didn't know that already.

Here are the parts that went into our long-block, minus the carb, intake and distributor. The 383 stroker kit from Racing Head Service was based on a SCAT cast crank and also included cast pistons, Speed Pro rings, reconditioned 5.7-inch rods with ARP bolts, Clevite 77 bearings, and a flexplate. We ordered the balance of parts from Summit Racing Equipment, including a Comp Cams Xtreme Energy cam and lifter set, pushrods, and Magnum roller rockers; HPX freeze plugs; ARP hardware; and a Summit oil pump, timing set, and oil pan.

Jim Grubbs Motorsports (JGM) performed all of our machine work and guided us through assembly; the block was counterbored for ARP main cap studs, which were more of a luxury than a necessity on this street mill. JGM also cleaned the threads with a bottoming tap to cut each hole to the same depth so the studs would seat evenly.

We stabbed in a 'stick from Comp Cams' Xtreme Energy series (PN XE256H) with 212/218 degrees of duration at 0.050-inch tappet lift, 0.447/0.454 lift, and a 110-degree lobe separation angle. We didn't need a small base circle cam, but we did check to make sure none of the lobes contacted the connecting rods, which can be a problem on strokers. This grind produces sick amounts of torque down low but pretty much runs out of breath above 5,000 rpm.

In a bid for a little more top-end power, we shelled out an extra $150 to have Bill Mitchell bowl-blend the intake ports before shipping the heads.

According to Pete Christensen at JGM, bolting an oil pump in place can alter the number-five main-bore diameter by up to 0.001 inch, so this torque plate is installed whenever a block is align-honed at the shop.

During short-block assembly, we checked ring endgap with a feeler gauge to make sure it was at least 0.016 inch for the moly top rings and 0.012 for the ductile iron second rings.

JGM disassembled and hot-tanked the Torquers and then milled 0.004 inch off each deck surface to clean them up. We didn't cc the heads afterward, but the cut probably knocked about 1cc off the 76cc chambers, which won't have much of an effect on our calculated CR.

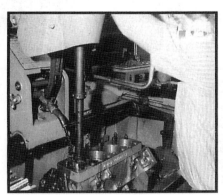
After the block was bored 0.030-over to clean up the ridge wear, Christensen honed each cylinder to its final dimension using a torque plate.

Instead of spending good money to rebuild junk smog heads, we ordered up a set of World Products' cast-iron S/R Torquer heads with 76cc chambers and 170cc intake runners from Bill Mitchell Hardcore Racing Products. For a street price of around $700, they come assembled with Manley stainless steel 2.02/1.60 valves, valvesprings, and retainers, and flow better than any production iron head.

Prior to decking, JGM mocked up the engine with pistons at each corner and measured the deck height, which varied from 0.028 to 0.031 inch in the hole (this is the distance from the piston deck to the cylinder-head deck surface). To achieve our target 9.6:1 compression with 76cc-chamber World Products S/R Torquer heads, 4cc piston valve reliefs, and head gaskets with 0.041-inch compressed thickness, the block was milled until the slugs were 0.010 inch in the hole.

After applying thread sealer, we threaded in the ARP studs hand-tight, then sealed up the short-block with Fel-Pro (PN 1003) head gaskets (0.041-inch compressed thickness). We torqued the nuts in sequence to 65 lb-ft using moly lube.

What It Cost

We tried to avoid gold-plating this buildup from the outset but spent more than we expected by the time every nut and bolt was counted. Costs escalated for a variety of reasons, including our desire for some trick parts. For one thing, we didn't really need to run head and main cap studs on an engine with this level of power. In fact, we're lucky that the head studs didn't leak on us, which is common with used blocks even when thread sealer is used.

We also spent a lot on an aftermarket billet distributor and an aluminum water pump. We got what we paid for, but we could have saved money without sacrificing power by using more OEM parts. For example, we would have saved about $400 by reusing all our hardware, including the head and main cap bolts. (If we could have found it all, that is.) Also, keep in mind that we actually added up all the costs, including the price of the core, machine work, and every part necessary to get the motor running on the dyno. The only things not included were headers and an alternator. If you're starting with even a few of these parts and spend selectively, you could easily shave well over $1,000 off this price.

Part	Source	Price
Core	Junkyard	$100.00
Stroker kit	RHS	699.00
World S/R Torquers	Bill Mitchell	725.00
Bowl blend	Bill Mitchell	150.00
Hot tank, bore and hone, machine for studs, align hone, deck block, and surface heads	JGM	600.00
ARP head stud kit	Summit	94.50
ARP main stud kit	Summit	45.95
ARP engine fastener	Summit	67.95
ARP cam bolt kit	Summit	3.00
ARP balancer bolt	Summit	17.50
Comp Cams Magnum roller-rocker arms	Summit	128.95
Comp Cams pushrods	Summit	35.69
Comp Cams Xtreme Energy cam kit	Summit	96.95
HPX freeze plugs	Summit	15.50
True Roller timing set	Summit	31.95
Oil pump w/ pickup	Summit	54.95
Oil pan	Summit	29.99
Chrome valve covers	Summit	22.95
Fel-Pro gasket set	Summit	89.95
Edelbrock Performer RPM manifold	Summit	117.00
MSD Pro Billet HEI distributor	Summit	389.00
MSD Super Conductor wires	Summit	59.00
Edelbrock Victor series water pump	Summit	149.95
Holley 4779 carb	Summit	318.00
Spark plugs	NAPA	7.92
Oil filter	Kragen	2.59
Fan belt	Kragen	5.99
6 quarts 30W oil	Kragen	8.10
Total		**$4,067.33**

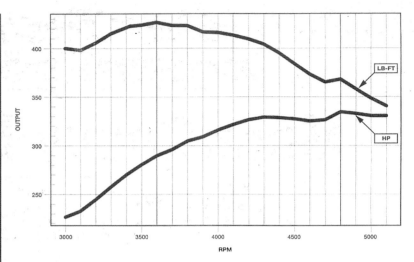

The valvetrain consists of 1.52:1 Comp Cams Magnum roller-tipped rocker arms and came with screw-in studs and guideplates that required using hardened pushrods from Comp Cams.

Dyno Results

RPM	HP	LB-FT
3,000	228.2	399.6
3,100	235.1	398.3
3,200	246.9	405.2
3,300	261.1	415.5
3,400	272.9	421.5
3,500	282.0	423.2
3,600	292.0	425.9
3,700	297.5	422.2
3,800	305.4	422.0
3,900	309.2	416.3
4,000	317.3	416.7
4,100	322.6	413.2
4,200	327.5	409.5
4,300	330.8	404.1
4,400	330.8	394.8
4,500	328.0	382.8
4,600	326.0	372.2
4,700	327.0	365.4
4,800	338.0	369.8
4,900	333.7	357.7
5,000	332.6	349.4
5,100	332.1	342.1

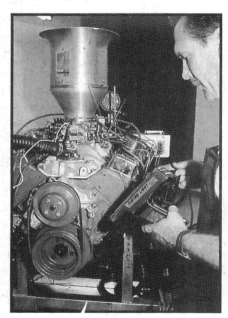

For engine startup and break-in on Westech Peformance Group's Super-Flow dyno, initial timing was set at 12 degrees BTDC (34 total), but fine-tuning determined that the engine liked 8 degrees of initial and 30 degrees total. This is good in an engine that may occasionally be used for towing because it greatly reduces the likelihood of detonation under load.

Out of the box, the Holley 750 double-pumper had 71 jets in the primary circuit and 80s on the secondary side, which ran lean. After experimenting with several richer stages, Westech's John Baechtel produced peak power with 74s up front and 81s in the back. **CC**